FPGAs and Programmable LSI:
A Designer's Handbook

FPGAs and Programmable LSI:
A Designer's Handbook

Geoff Bostock

Butterworth-Heinemann
Linacre House, Jordan Hill, Oxford OX2 8DP
A division of Reed Educational and Professional Publishing Ltd

R A member of the Reed Elsevier plc group

OXFORD   BOSTON   JOHANNESBURG
MELBOURNE   NEW DELHI   SINGAPORE

First Published 1996

© Geoff Bostock

All rights reserved. No part of this publication may be
reproduced in any material form (including photocopying
or storing in any medium by electronic means and whether
or not transiently or incidentally to some other use of
this publication) without the written permission of the
copyright holder except in accordance with the provisions
of the Copyright, Designs and Patents Act 1988 or under
the terms of a licence issued by the Copyright Licensing
Agency Ltd, 90 Tottenham Court Rd, London, England W1P 9HE.
Applications for the copyright holder's written permission
to reproduce any part of this publication should be
addressed to the publishers.

British Library Cataloguing in Publication Data
A catalogue record for this book is available from the British Library.

ISBN 0 7506 2883 9

Library of Congress Cataloguing in Publication Data
A catalogue record for this book is available from the Library of Congress.

621.3819'5835

BOS

Composition by Genesis Typesetting, Laser Quay, Rochester, Kent
Printed in England by Clays Ltd, St Ives plc

Contents

Preface   vii

1. Classic Logic Devices   1
    1.1. Standard logic families   1
    1.2. Application-specific integrated circuits   11
    1.3. Programmable logic devices   17

2. Programmable LSI Techniques   30
    2.1. LSI PAL devices   30
    2.2. Logic cell arrays   34
    2.3. Antifuse FPGAs   38

3. Designing FPGAs   41
    3.1. Introduction   41
    3.2. Logic equations   42
    3.3. State equations   48
    3.4. Hardware description languages   50
    3.5. Schematic capture   55
    3.6. Conclusion   66

4. Large PAL Structures   67
    4.1. MACH families   67
    4.2. MAX families   76
    4.3. XC7000 EPLDs   85
    4.4. FLASHlogic   90
    4.5. (is)pLSI families   93
    4.6. LSI sequencers   97
    4.7. FLASH370 series   101

5. RAM-Based FPGAs 105
5.1. LCA families 105
5.2. Atmel FPGAs 116
5.3. FLEX8000/10K families 120
5.4. Other RAM-based FPGA families 125

6. Antifuse FPGAs 130
6.1. Actel families 130
6.2. QuickLogic/pASIC380 FPGAs 139
6.3. Xilinx XC8100 FPGAs 142

7. Selecting and using FPGAs 146
7.1. Basic criteria 146
7.2. FPGA testing 158
7.3. Programming FPGAs 165
7.4. Migration to ASIC 169

8. Using FPGAs 172
8.1. Design techniques 172
8.2. Board-level considerations 183
8.3. Conclusions 186

Appendix A. Device manufacturers 189

Appendix B. CAD and programmer suppliers 200

Appendix C. References 206

Appendix D. Trade Marks 207

Glossary 209

Index 211

Preface

At first sight this book may seem to be a précis of the manufacturers' data books and, to a certain extent, this is true. I believe that this does not make it any less worthwhile. The pile of data books which has been condensed into three of the chapters is about half a metre high. If I have managed to extract the relevant points and summarize them in less than half a book that will have saved the reader much time and effort.

The book is more than a précis of data, however. The intention, also, is to provide the newcomer to programmable logic a complete picture of all that is involved in taking a concept to a piece of hardware. It covers the options available in both software and hardware and should prove a major influence in guiding him or her to the best decision regarding design method and target device

As well as just reading application notes and data, a major source of inspiration has been the experience gained in designing FPGAs (field programmable gate arrays) myself. While it is not possible to pass on all experience by word of mouth or pen, I have tried to pass on as much as possible, including pitfalls to avoid.

As I explain in the book, FPGAs are expanding to fill a void being left by masked ASICs (application-specific integrated circuits) caused by the reduction of geometry sizes, and consequent increase in gate counts. Furthermore, designs which previously used multiple PLDs (programmable logic devices) can now take advantage of this technology. As they purloin market share from above and below they are clearly going to increase in importance during the next few years – hence the need for a book like this.

Geoff Bostock

Acknowledgements

Figures 4.18, 4.19, 4.20, 4.21, 4.22, 5.1, 5.2, 5.3, 5.4, 5.5, 5.6, 5.7, 5.8, 5.9, 6.11 courtesy of Xilinx, Inc. © Xilinx, Inc. 1994, 1995. All rights reserved.

1 Classic logic devices

1.1 Standard logic families

1.1.1 Gate functions

The primary building block of logic circuits is the **logic gate**. This a device which operates on two or more logic signals to give an output which is defined by a logic operator. The standard logic operators are AND, OR and INVERT (which only acts on one signal).

The two classical logic families are TTL, which uses a nominal 5 V supply, and 4000 series CMOS, which can work from a supply of between 3 V and 15 V, although 4000 series has been largely superseded by the HC family. A logic LOW signal is usually defined as being close to 0 V or GND (ground); a logic HIGH signal normally sits close to the supply voltage (Vcc), or at least above half Vcc.

An AND function is defined as only giving a HIGH output when all the inputs are HIGH; the OR function has a HIGH output when any one of its inputs is HIGH. The INVERT function changes a logic signal from HIGH to LOW, or vice versa. These functions may be written down as logic equations as follows:

AND function: $Y = A \star B \star C$ or $Y = A \& B \& C$
OR function: $Y = A + B + C$ or $Y = A \# B \# C$
INVERT function: $Y = /A$ or $Y = !A$

The alternative notations exist because different logic compilers have adopted different conventions. Using '$\star$', '+' and '/' for logic clashes with their more familiar use as arithmetic operators in programming languages so, for the remainder of the book, we will use '&', '#' and '!'. Figure 1.1 shows an alternative way of showing logic relationships, the truth table. The symbol 'H' represents a HIGH, sometimes replaced by '1', 'L' stands for

A	B	C	Y
1	1	1	H
0	X	X	L
X	0	X	L
X	X	0	L

AND

A	B	C	Y
1	X	X	H
X	1	X	H
X	X	1	H
0	0	0	L

OR

A	Y
1	H
0	L

INVERT

Figure 1.1 Logic function truth tables

LOW, with '0' as an alternative, while 'X' means don't care – either HIGH or LOW.

These three operators form the basis of all possible logic circuits. For example, the exclusive-OR gate has a LOW output if the two inputs are the same but HIGH if they are different. Its logic operator is written as:

Y = A $ B or, sometimes, Y = A :+: B.

It is logically equivalent to:

Y = A & !B # !A & B

These logic functions may also be represented diagrammatically. The standard gate symbols are shown in Figure 1.2; as well as AND, OR and INVERT, NAND and NOR gates are also depicted. NAND is an

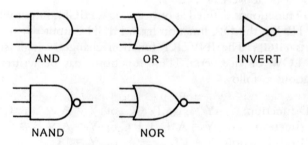

Figure 1.2 Standard gate symbols

AND followed by an INVERT, NOR is OR followed by INVERT; a small bubble on an output, or an input, signifies an inversion. The exclusive-OR symbol, and its equivalent logic circuit, are drawn in Figure 1.3.

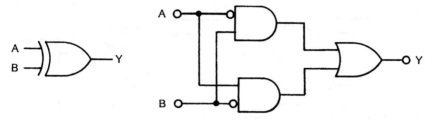

Figure 1.3 Exclusive-OR symbol and equivalent circuit

1.1.2 Sequential logic

The output of a gate does not depend on the order in which the signals are applied. If both inputs of a two-input AND gate are LOW, the output will also be LOW; if A goes HIGH before B, or if B goes HIGH before A the result will be the same – two HIGHs on the inputs yield a HIGH on the output.

Consider the circuit of Figure 1.4. If input LE is HIGH the output, Q, will be the same as input D. This is because the output of the lower AND gate is LOW, irrespective of D. Suppose that LE is now taken LOW; the

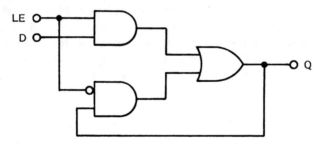

Figure 1.4 D-latch circuit

upper gate now has a LOW output so Q will follow the output of the lower gate. If D was HIGH when LE went LOW, Q was also HIGH, so the lower gate was HIGH and Q will stay HIGH. Conversely, if D was LOW, Q will stay LOW.

This function is known as a **latch**. While LE is HIGH, the latch is transparent and Q follows D; when LE is LOW, the level of D when LE went LOW is latched into Q. The output of the circuit depends on the sequence in which the signals are applied, hence the term **sequential circuit**.

Figure 1.5 shows two latches in series, with the LE signal inverted between them. When the first latch is transparent, the second is latched and vice versa. A signal applied to D1 will appear at D2 while LE1 is HIGH, but

4   FPGAs and Programmable LSI: A Designer's Handbook

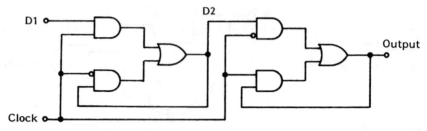

Figure 1.5   *Master-slave D-type flip-flop*

will not be transmitted to Q until CLK goes LOW, sending LE2 HIGH. At this time LE1 will go LOW and lock out any changes on D1. It appears that the signal on D1 is sent through to Q as CLK changes from HIGH to LOW. This is the principle of the **master-slave flip-flop** or **D-type flip-flop**.

1.1.3 Practical logic circuits

Devices containing one or more gates, latches or flip-flops form the basis of the standard logic families. Circuit designers can use these integrated circuits to build up more complex functions by interconnecting these SSI (small-scale integration) on a printed circuit board. However, the device manufacturers anticipated these requirements by producing MSI (medium-scale integration) parts which contain many of the standard circuit functions which can be built from gates and flip-flops.

A typical combinatorial MSI function is a one-to-eight line **decoder/demultiplexer**. The circuit diagram for this function is shown

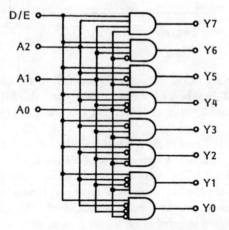

Figure 1.6   *One-to-eight line decoder/demultiplexer*

Table 1.1 Truth table for line decoder/demultiplexer, Figure 1.6

D/E	A2	A1	A0	O7	O6	O5	O4	O3	O2	O1	O0
0	X	X	X	L	L	L	L	L	L	L	L
1	0	0	0	L	L	L	L	L	L	L	H
1	0	0	1	L	L	L	L	L	L	H	L
1	0	1	0	L	L	L	L	L	H	L	L
1	0	1	1	L	L	L	L	H	L	L	L
1	1	0	0	L	L	L	H	L	L	L	L
1	1	0	1	L	L	H	L	L	L	L	L
1	1	1	0	L	H	L	L	L	L	L	L
1	1	1	1	H	L	L	L	L	L	L	L

in Figure 1.6. The logic levels on the three address inputs represent a binary number in the range 000 (0) to 111 (7). The logic level on the data/enable input is transmitted to the output selected by the input address. The **truth table** for this function is shown in Table 1.1.

Similarly, MSI functions can be built from sequential elements. Figure 1.7 illustrates this point with the circuit of a 4-bit shift register. Data on the inputs is loaded into the flip-flops when the shift/load signal is LOW and there is a LOW to HIGH clock transition. When shift/load is HIGH, the data is moved one place to the right on a clock edge. This circuit can form

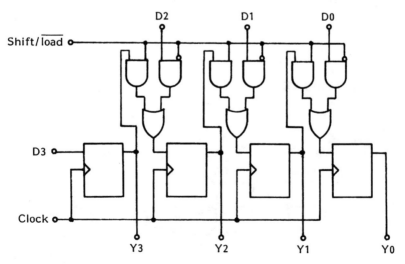

Figure 1.7 Four-bit shift register

6   FPGAs and Programmable LSI: A Designer's Handbook

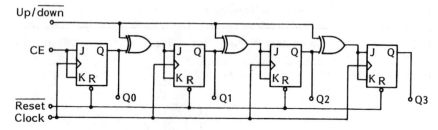

Figure 1.8   *Four-bit counter*

the basis of some arithmetic functions, or it may be used in communications to change the data format from parallel to serial or vice versa.

A different type of flip-flop, the **J-K flip-flop**, is used in the counter circuit in Figure 1.8. A J-K flip-flop behaves like a D-type when the J and

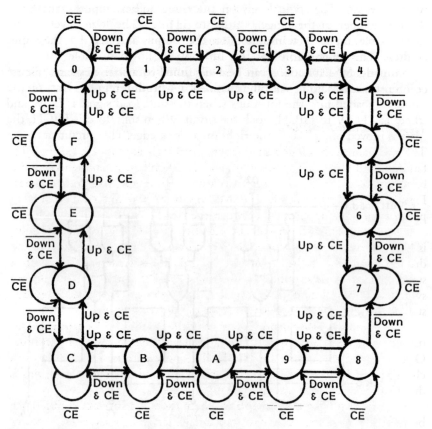

Figure 1.9   *Four-bit counter state diagram*

K inputs are complementary, but if both are HIGH the output changes from HIGH to LOW or LOW to HIGH, the so-called **toggle function**. If both are LOW, no change occurs on a clock edge; the output data is unchanged, the hold condition. As long as the CE (count enable) and reset inputs are held HIGH, the **counter** function will increment by one on each clock transition.

Taking the reset LOW makes every output LOW, resetting the counter to binary 0000. Reset is an asynchronous function which is built into some flip-flops; it operates independently from the clock and allows the flip-flop to be put into a known condition. Some flip-flops also include an asynchronous set input which puts the output HIGH.

Operation of the counter may be described by a **state diagram**. Each combination of output levels corresponds to a binary number from 0000 (0) to 1111 (15), and represents a distinct state of the counter. Figure 1.9 shows the counter state diagram. Each state is represented by a circle containing the output levels in that state. The arrows are labelled with the logic inputs which enable the jumps between states. A recursive arrow means that there is no jump and gives the hold condition for that state.

The CE must be TRUE for counting to proceed; if CE is FALSE the counter holds its present state.

Many sequential functions are best described by a state diagram, just as combinatorial functions are often defined in a truth table.

1.1.4 Physical structure

The two main technologies used for standard logic are bipolar, for the TTL families, and CMOS, for the 4000 and HC families. Examination of the basic circuit diagrams of a standard gate built in the two technologies, Figures 1.10 (TTL) and 1.11 (CMOS), shows the main differences in performance and application between the two.

TTL always has a direct current path from Vcc to 0 V; when any input is LOW current will flow through R1 and via the output transistors of the driving stages. If all the inputs are HIGH, T2 is switched ON with a standing current determined by the values of R2 and R3. Also, during switching, both output transistors conduct momentarily causing a substantial current spike to be drawn from the power supply.

CMOS, on the other hand, always has either the lower n-channel ladder turned OFF, when any input is LOW, or all the upper p-channel transistors OFF when all the inputs are HIGH. The only current which flows is a charging/discharging current when any of the nodes changes level; this is due to the capacitance associated with any node.

In terms of **power consumption**, then, there is a substantial difference between the two approaches to standard logic. TTL is a relatively power hungry technology, consuming anything from less than 1 mA to over 6 mA,

8 FPGAs and Programmable LSI: A Designer's Handbook

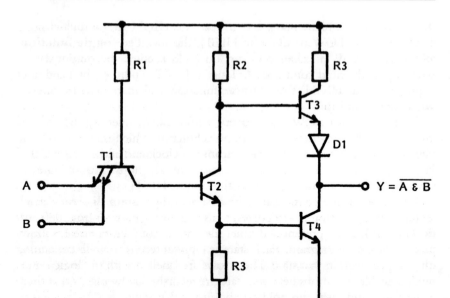

Figure 1.10 *TTL NAND gate*

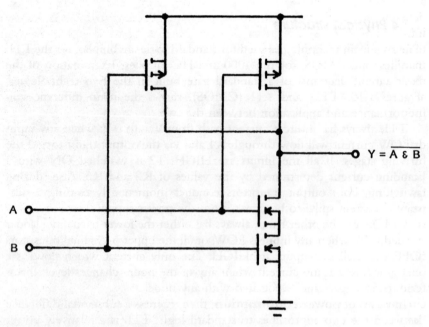

Figure 1.11 *CMOS NAND gate*

depending on the resistor values which are chosen according to the speed of operation required for the particular TTL family. Current consumption also increases with operating frequency because of the current spike which we noted at every switch of the outputs.

When no inputs are being switched in a CMOS gate, the only current which flows will be leakage current through the OFF MOS transistors. CMOS, therefore, consumes virtually zero power at zero frequency. As with TTL, the supply current increases with operating frequency as both internal and external capacitances are charged and discharged more often. In summary, CMOS is a low-power technology while TTL is a high-power technology.

The other major parameter which is usually important to circuit designers is **operating speed**. A crude measure of this is the **propagation delay** between the inputs and outputs of a logic gate. The most important part of this delay is the time taken to charge and discharge the node capacitances inside and outside the circuit. Another component is the time needed to remove stored charge from transistors in the circuit.

The time for a change in voltage is $(C \times V)/I$ where C is the node capacitance, V is the voltage change and I is the charge/discharge current. A low delay time is thus achieved by having a low capacitance and voltage change, but a high current.

Internal capacitance depends largely on feature size which is likely to be a common factor between both TTL and CMOS. External capacitance depends only on the packaging of the integrated circuit and the PCB layout into which it is inserted, both of which should be independent of the technology. The voltage change for TTL is 2.0 V − 0.8 V, allowing for noise margins for most families. At 5 V, CMOS noise margins dictate a voltage change of 3.5 V − 1.5 V, slightly more than for TTL but the actual voltages at which switching occurs are not defined exactly so there will probably be little significant difference between the two.

The current proves to be the real point of divergence. TTL families are driven by transistors which are saturated, or held on the verge of saturation by Schottky diodes, while CMOS current sources are more nearly resistive. For a given scale of technology, then, TTL families are traditionally the faster. Differences in the geometry of bipolar and CMOS transistors have made CMOS transistors easier to scale down in size so the current situation is that CMOS devices can be as fast as TTL.

Another aspect of device speed is the output **slew rate**. If the rise or fall time of an output is comparable with the physical delay along the PCB tracks, reflections can become a problem and outputs may need to be terminated. The faster TTL families could fall into this category in some circumstances while this was not usually a problem with the older CMOS devices. Now, however, with smaller geometries and faster edges, CMOS can need the same remedial treatment in some circumstances.

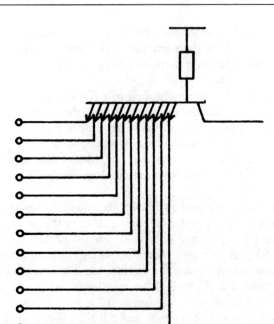

Figure 1.12 *TTL wide gating input structure*

One other important difference between the two technologies lies in the input structures. A TTL input is a compact diode structure and there is virtually no limit to the number of inputs which can be attached to a single gate, as can be seen in Figure 1.12. In a CMOS gate transistors must be stacked up, one transistor for each input. If there are too many transistors in the stack, the voltage drop across individual transistors becomes too small for correct operation and the gate fails to work. The consequence of this is that, for example, the twelve-input gate in Figure 1.13 must be made by using two stages of gating where one stage is possible in TTL form.

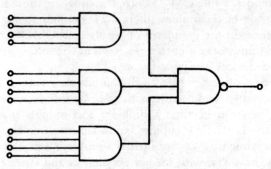

Figure 1.13 *Twelve-input CMOS gate*

1.1.5 Large-scale integration

As processes improved to the point where a thousand or more transistors could be laid on a single chip, LSI (large-scale integration) became feasible. The situation is different from MSI. MSI functions can still be looked on as building blocks with universal application; for example, a four-bit counter might be used in a computer, a CD player or a digital multimeter. LSI circuits are usually a self-contained function, the most prolific example being the microprocessor. Apart from the microprocessor, most LSI functions are specific to a particular application; for example, a UART will normally only be found in communications equipment and a frequency synthesizer in tuners.

At first it was thought that microprocessors would drastically reduce the volume of SSI and MSI being used, but two effects made the reverse true. Although microprocessors operate at frequencies in excess of 100 MHz, a simple logic function, such as ANDing two bytes, may need several operations to acquire the data, perform the function and then provide an output. The total cycle may occupy over 100 ns, compared with less than 10 ns in readily available logic chips. The processor is also prevented from performing other tasks during this period so it makes sense to continue with hard-wired logic or SSI for simple logic functions.

The other effect is the need to interface the microprocessor to the outside world. Nearly every application, for example, needs an address decoder so that data and process instructions can be routed to and from the processor. This, and the need to customize many of the general-purpose peripheral circuits, adds to the number of discrete logic circuits surrounding the microprocessor.

In principle, these added chips can be combined into a single chip. Economies of scale dictate that this is not a practical approach unless the circuit is going to be used in upwards of a hundred thousand. These custom circuits are used in some applications where the quantities allow the cost of designing the chips, making masks, designing test sequences, and so on, to be amortized over a sufficiently large number.

Most applications are not large enough to benefit from this approach, but we can examine other ways in which the circuit designer can condense his hard logic into a small number of LSI chips.

1.2 Application-specific integrated circuits

1.2.1 Physical structure

The standard process for manufacturing integrated circuits involves growing a layer of silicon dioxide on a silicon wafer, etching windows in the surface layer and then introducing a controlled amount of impurity into the windows by vapour or electron beam deposition. Successive layers of different impurities laid down through different sized and shaped windows build up the active components in the surface layer of silicon.

Connections between the components are made by evaporating a conductor, usually aluminium or polycrystalline silicon, over the silicon dioxide. Previously etched windows allow contact to be made to the desired components. The conductor is then etched into tracks to define the circuit connections. Two or more conductor layers can be used by sputtering layers of silicon dioxide between the conductor layers.

In a standard or custom LSI circuit the layout is made by placing components on a 'floor plan' according to the schematic circuit diagram which defines the function of the chip. It will be laid out to minimize the area of the finished circuit, bearing in mind the design rules for the process. Usually, it is desirable to make the chip with a particular aspect ratio, often square, to aid assembly. Also, components which are close on the schematic will need to be close on the chip to make the conductor tracks as short as possible.

Clearly the arrangement of components is suitable for only the particular circuit under consideration; if any changes need to be made, or a new

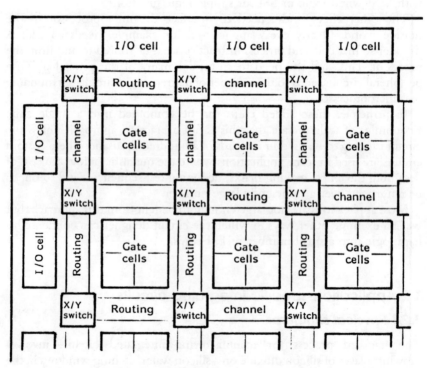

Figure 1.14 *Channel routing ASIC*

circuit laid out, the component positions will need to be changed as well. However in a **gate array**, which is the simplest type of masked ASIC (application-specific integrated circuit), the components are laid out in a predetermined pattern and the conducting layers tailored to the circuit schematic diagram.

Usually, the circuit components are set on a rectangular grid. Connections between components may be along 'routing channels' in the gaps between components or, if no gaps are left, by routing across unused components. Figures 1.14 and 1.15 show the channel routing and 'sea of gates' approaches, respectively.

The other choice to be made in a gate array is what the basic component should be. The simplest component is a two-input NAND or NOR gate. In principle, any logic circuit can be built from basic gates; we have already seen, in Figures 1.3 and 1.5 how an exclusive-OR gate and a D-type flip-flop may be constructed. On the other hand, using two-input gates to build a sixteen-input composite gate, which may be needed to decode a microprocessor address, would take five levels of logic plus at least fifteen gates. This would result in a long propagation delay and use a significant proportion of the gate array resources.

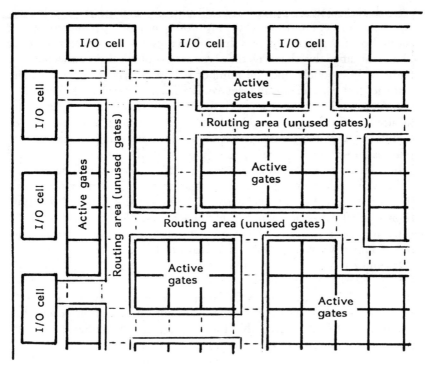

Figure 1.15 *Sea-of-gates ASIC*

An alternative approach is to use a more complex cell. One example is a cell containing four p-channel and four n-channel transistors. The cell can be configured in several different ways; a four-input gate, a three-input gate plus invertor and two two-input gates are examples. The transistors can also be made into transmission gates which form the basis of flip-flops.

Gate arrays exist with a variety of cell designs and interconnection methods, but the ways in which designs are entered are, on the whole, quite similar.

1.2.2 Designing ASICs

Designing a logic circuit is virtually independent of the physical form in which the circuit will be implemented. There are, as we shall see later, differences in the way in which tri-states and state machines, for example, are designed in different end products, but multiplexers, counters and other standard logic functions may be used to build up any logic system.

In effect, the designer is presented with a library of building blocks which he connects together to produce the desired result. If he is using TTL or CMOS standard logic the library will be listed in a data book which presents the relevant features of each device. These include the pin-out to show which functions appear on which pin, DC parameters which show how each device interfaces with any other and AC parameters which indicate how fast signals will pass through the system. The AC performance will be modified by the way the devices are physically connected; the lengths of PCB tracks and the number of inputs driven by each output affect the load capacitance and, hence, the delay through each chip.

An ASIC designer will also work from a library. In this case the library may also take the form of a data book but the information will be presented in a different way. For a start there is no need for pin-out information. The circuit is most probably being drawn on a CAD system and standard symbols for gates, flip-flops and more complex structures will be used.

The ASIC equivalent of a standard logic function is a **macro**. A macro is one or more cells with the wiring to produce the required function. When a signal path joins two cells in a macro, the physical location of the electrical nodes is built into the macro description so the designer does not need to know where the connections have to be made on the chip.

DC and AC parameters are quite significant in ASIC design. Standard logic functions have built-in buffers which ensure that all devices in a given family will drive each other with compatible voltage levels and current drains. These buffers also guarantee a maximum propagation delay when driving a specified load. In ASICs there are no internal buffers so fan-out becomes an important factor when making a design.

Simulation is a very important part of the ASIC design process. This is not only to ensure that the logic function of the finished device meets the original goal, but to check that the design rules have not been violated. Part of the skill lies in ensuring that all parts of the design are testable; that every part of the circuit can be exercised by applying signals to the various inputs, and that the results can be seen by changes at the outputs. Bed-of-nails testing is not feasible for an ASIC and simulation will show any deficiencies in device testability.

ASIC macros are usually more 'fine grained' than standard logic circuits. To use an invertor in TTL or CMOS means either specifying a chip with six invertors or wiring a NAND or NOR gate as an invertor, if one is conveniently spare. This may mean wasting five invertors or, if they can be used elsewhere in the circuit, running long tracks to and from another part of the PCB. Similarly, flip-flops usually come in pairs, counters in four-stage blocks and so on.

These restrictions do not apply to ASICs; an invertor will, at worst, take a single cell and may be able to be included in a cell which includes another function. Likewise, a counter with five, six or seven stages can be included without wasting any cells; moreover, features such as fast lookahead carry can be designed in to give a better performance than might be possible with discrete logic chips.

The downside is apparent when the whole design is considered. Gate arrays are made with a fixed number of cells so a design will have to be built in the device with the next higher cell count than the actual number needed. If the smallest array has 1000 cells, the next highest 1500 cells and a certain design needs 1010 cells, over 30% of the chip will be wasted.

Having selected a suitable array, the next stage is to map the logic diagram onto the physical cells. This is now normally automated; the process being referred to as place-and-route. The place function involves assigning cells to each of the macros in the design, routing being the connection of the cells according to the connections between the macros already specified. With a channel architecture it is usually possible to use 95% or more of the cells, depending on the number of connections allowed in each routing channel. The sea-of-gates type of array is often limited to about 60% utilization as connections have to be made over unused cells. These usually give shorter track lengths, on the whole, as the connections can be made by more direct paths.

The final stage in the design process is post-layout simulation. The first simulation will give an idea of the delays and timing performance of the circuit, by including fan-out and a nominal delay for connections. Until the routing has been completed, though, an accurate measure of the delays cannot be obtained. The post-layout simulation includes an estimate of the extra delays due to the actual track lengths and should give an accurate picture of the performance of the final device. If the performance does not measure up to the requirements

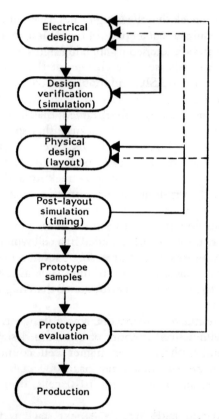

Figure 1.16 *ASIC design flow*

there is usually an opportunity to make manual changes to the layout to reduce the delay in critical paths.

A typical design flow is shown in Figure 1.16.

1.2.3 Using ASICs

Once a design is finalized it will have to go to the manufacturing stage. ASICs are semi-customized integrated circuits; they start processing as a standard component, because all the diffused components are independent of the final design. The final stages are, however, customized; each different design is application-specific and requires its own pattern to be imposed on the upper levels of the chip. A stock of partly finished wafers can be held in readiness for customization.

Customization is achieved by making masks to create the connections already specified by the place-and-route step in the design process. Four or more masks may be required; two to define the interconnection tracks to be etched into the two track metallization layers and two to allow contact holes to the silicon and between the two metal layers to be made.

Mask making and processing are both time consuming and costly. Frequently these activities take place in a different country from that in which the design is done and the finished chip may be packaged in a third country. All is well if the final device performs as expected but, whether because of a mistake in design or an imperfect specification, if changes have to be made, much delay and expense can result.

The NRE (non-recurring engineering) costs involved in masked ASICs make them more suitable for projects with expectations of long, stable high-volume runs. As with all aspects of design and engineering, it is sometimes expedient to make trade-offs. The savings in packaging plus the need for high performance in a confined space might make it economic to use a gate array for an expected run of only a few hundred, but it is more usual to be looking at a volume in excess of ten thousand for gate array designs.

It seems from the analysis so far that circuit designers are stuck with only two choices for building logic circuits, standard families and masked ASICs. Fortunately, technology has evolved a third option, the **programmable switch**, and this is the basis of the rest of this book.

1.3 Programmable logic devices

1.3.1 Programmable switches

There are four types of programmable switch which have been used in any volume on programmable logic devices, and we can describe them briefly here.

The simplest form of AND gate is a diode array, shown in Figure 1.17. If all the inputs are near V+ none of the diodes conduct and the output will also be pulled up to near V+. Any input taken to 0 V will pull the output to a diode drop above 0 V. In a standard logic circuit all the inputs are available at device pins, but in a PLD (programmable logic device) a programmable switch is placed in series with the input so that the user can select which signals will affect the gate.

A metal fuse was the first type of switch and is traditionally associated with bipolar PLDs. An alloy, such as nichrome or tungsten-titanium, is evaporated onto the surface of the chip and etched into small strips about 5 μm wide and 20 μm long. A current pulse of about 50 mA is sufficient to vaporize the metal, which fuses into the overlying silicon dioxide leaving an open circuit at the fuse site.

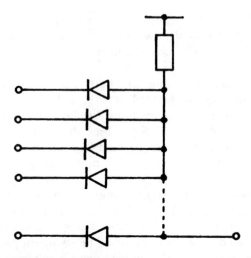

Figure 1.17 Diode array (AND gate)

An alternative fuse in bipolar technology is the AIM (avalanche-induced migration) device. This is a small transistor with a floating base, so that the emitter-collector path is normally high impedance. If the emitter-base junction is deliberately overstressed, the aluminium from the emitter contact will migrate into the junction causing a short-circuit. The emitter-collector path is now a diode and can be used in its own right as a gating element; in this case, then, the fusing process is used to establish the required inputs to the gate.

In MOS technology the transistors are, themselves, very efficient switches which can be turned on and off by applying HIGH or LOW signals to their gates. By adding a second gate, floating between the control gate and the conducting channel, the transistor threshold can be varied by charging or discharging the second gate. In the low threshold condition the transistor acts normally, but in the high threshold state the channel is held off permanently. The floating gate can be charged electrically but needs ultra-violet light to discharge it. By adding silicon nitride to the sandwich, the surface states can be defined for positive and negative charges and the floating gate can be predictably discharged electrically. The transistors themselves are used for logical gating, as in Figure 1.18.

A later development is the antifuse. This is simply a thin layer of silicon oxide/nitride sandwiched between two conducting layers, which may be either silicon or metal. A short voltage pulse of 15 V – 20 V ruptures the insulating layer and the heat alloys the two layers together. A resistor of less

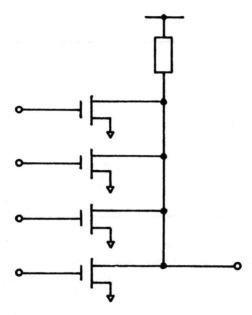

Figure 1.18 *MOS transistor array*

than 1 kΩ results, sufficiently low to appear as an ON switch to signals in a CMOS environment.

1.3.2 PAL devices

The most commonly used PLDs are PALs (programmable array logic). Although not the first PLDs they are the easiest to use and took the largest share of the programmable logic market.

They are based on the idea that any combinational logic function can be represented by a 'sum of products' equation. AND functions are sometimes referred to as product terms, by analogy between logic equations and arithmetic equations, and OR functions as sum terms; sum of products means just the OR combination of a number of AND terms. The justification for this concept is the **Karnaugh map**.

A Karnaugh map, or K-map, is constructed by taking all the inputs to a given function and drawing a grid containing all the possible combinations of HIGH and LOW for those inputs. Conventionally, each axis of the grid is numbered with Gray code with half the inputs expanded along the x-axis and the other half in the y-direction. We can illustrate this by examining the truth table and K-map for an adder; the truth table is shown in Table 1.2.

Table 1.2 *Truth table for PAL adder*

A	B	CI	S	CO
L	L	L	0	0
L	L	H	1	0
L	H	L	1	0
L	H	H	0	1
H	L	L	1	0
H	L	H	0	1
H	H	L	0	1
H	H	H	1	1

The K-maps for S and CO are shown in Figure 1.19; each needs four AND terms on first inspection, but consider the two cells circled together in the CO map. These represent the AND-terms:

!A & B & CI
A & B & CI

which simplifies to just B & CI, because input A can take either sense and, therefore, be eliminated from the equation. In terms of logic analysis, this is because !A # A = 1.

The full equations for the adder are, therefore:

S = !A & !B & CI # !A & B & !CI # A & B & CI # A & !B & !CI
CO = B & CI # A & CI # A & B

In discrete logic these functions can be built as in Figure 1.20.

The structure of a simple combinational PAL is very similar to this circuit. Figure 1.21 shows how this same function could be incorporated into an imaginary PAL4H2. PAL numbering is quite logical; in our example the '4' refers to the number of inputs to the AND array, 'H' means

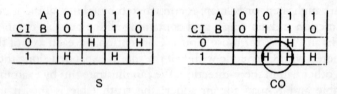

Figure 1.19 *Full adder Karnaugh maps*

Classic logic devices 21

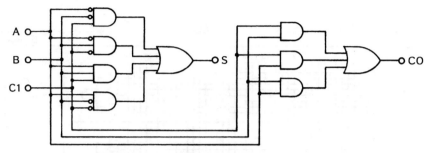

Figure 1.20 Discrete logic full adder

that the outputs are HIGH when one of the AND terms is true (active-HIGH) and '2' is the number of outputs from the PAL. The eight vertical lines in the figure carry the buffered/inverted signals from the four inputs – that is A, !A, B, !B, CI, !CI, D, !D, although D and !D are not used in this example. Each crossing point between a vertical signal line and the input line (which is really an eight-bit bus in this case) into each AND gate has

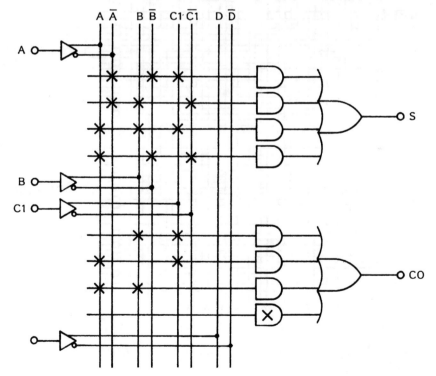

Figure 1.21 PAL4H2 full adder

22 FPGAs and Programmable LSI: A Designer's Handbook

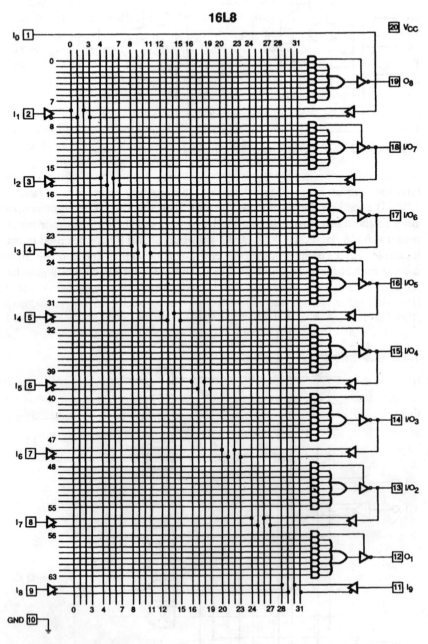

Figure 1.22 PAL16L8 circuit (reproduced by permission of Advanced Micro Devices)

a programmable switch which determines whether or not the signal is connected to the AND gate. The diagonal crosses indicate those fuses which are left intact for this application.

The simplest combinational PAL currently in production is the PAL16L8 whose circuit is shown in Figure 1.22. As with the first example, the numbering means that this PAL has 16 array inputs although, as six of these are fed back from outputs, the device has just ten dedicated inputs; in addition there are eight active-LOW outputs. Because of the feedback any configuration from ten inputs/eight outputs to sixteen inputs/two outputs can fit in to this PAL. Alternatively, some of the feedback pins can be used to make functions such as latches, or to augment the seven product terms per output for more complex functions.

By adding flip-flops to the outputs it is possible to make PALs capable of containing sequential functions. The PAL16R8 has all eight outputs registered and fed back to the AND-array, hence there are just eight direct inputs plus a common clock for the flip-flops and a tri-state enable to allow the outputs to be connected to a bus. The range of registered PALs also includes one with four flip-flops and four combinational outputs (PAL16R4) and a device with six flip-flops and two direct outputs (PAL16R6).

1.3.3 Generic PLDs, FPLAs and FPLSs

It is a drawback to the range of PALs described in the previous section that the designer is restricted to fixed architectures with eight, six, four or no flip-flops. The introduction of generic macrocells made PLDs more flexible

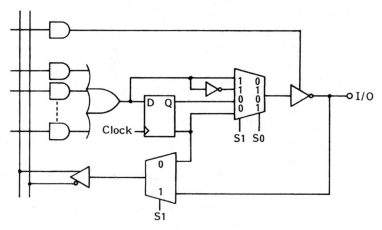

Figure 1.23 *PALCE22V10 macrocell*

architecturally. Figure 1.23 shows a typical **macrocell**; the flexibility is achieved by the use of **programmable multiplexers** to route the output signal through different paths.

The design of the macrocell varies slightly from device to device; our example is actually used in the 22V10 device. There are four possible sources for the output signal. These are combinational, combinational but inverted, registered and registered/inverted. The same programmable bit that selects the registered or combinational signal for the output also routes it back to the AND-array, but always inverted.

Generic PLDs can cover most programmable logic applications in a given package size, simply because of the range of input/output configurations which can be programmed into each device. The GAL16V8 (generic array logic with versatile outputs) can perform the same function as the PALs 16L8, 16R4, 16R6 and 16R8 plus innumerable other I/O combinations which are not found as standard PALs. The GAL16V8 is a 20-pin device and there is a 24-pin GAL20V8 which has similar features, but with four more inputs.

The most likely problem with fitting a logic circuit into a GAL16V8 or GAL20V8 is the number of product terms available. Combinational outputs have seven terms and registered outputs eight terms each and, if the function requires more terms, a different solution will have to be found. The first approach would be to try logic minimization, but most logic compilers incorporate a minimizer anyway. It might be worth trying to invert the output sense and minimize, but there is no guarantee that would be successful either.

The 24-pin GAL22V10 has more product terms, a variable number from eight to sixteen depending on which output is being used, and this will cope with most logic designs.

There is another class of PLD which can sometimes manage to overcome product term limitation problems with PALs and GALs; this the FPLA (field programmable logic array). The commonest FPLAs are the PLS153 and PLS173 which are 20 and 24-pin devices respectively, the PLS173 having four more inputs than the PLS153. Their difference is that they have 32 product terms which can be accessed by any output. This is achieved by having the OR gates connected by a programmable array to the AND gates, as can be seen in Figure 1.24.

All ten outputs are fed back to the AND-array, which therefore has eighteen buffered/inverted inputs altogether. As with the PAL circuit, there is a programmable switch at every crossing point between an input line and an AND gate input bus; unlike the PAL there is also a programmable switch at every crossing between an AND gate output and an OR gate input bus. This added versatility can give a higher logic content but there is a price to pay in terms of performance. The capacitance of the second array adds a significant extra delay to the

Classic logic devices 25

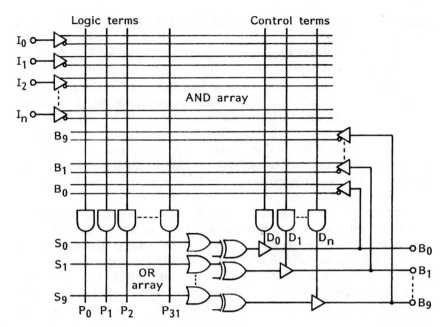

Figure 1.24 Programmable OR-array

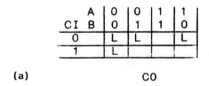

(a)   CO

INPUTS			OUTPUTS	
A	B	CI	S	CO
			[Active Level]	
			H	L
0	0	0	–	A
0	0	1	A	A
0	1	0	A	A
1	0	0	A	A
1	1	1	A	–

(b)

Figure 1.25 (a) Active LOW K-map for carry out; (b) minimized full adder truth table

signal path compared with a PAL or GAL made by the same process.

As an illustration of the product term saving possible with an FPLA we can take another look at the simple adder. If the K-map for CO is redrawn as an active-LOW function, as in Figure 1.25(a), we can see that three of the AND terms are identical to terms in the S output. We can see from the truth table in Figure 1.25(b), that only five product terms are used compared with seven in the PAL implementation. Thus, although FPLAs have fewer terms altogether than a PAL or GAL, judicious partitioning and use of the programmable output level can squeeze the logic into a smaller number of terms, while allowing a high term count if needed, on any particular output.

Just as registered PALs are formed by adding flip-flops to some or all of the outputs of a combinational PAL, so an FPLS (field programmable logic sequencer) is generated by adding flip-flops to an FPLA. Typically, the PLS155, PLS157 and PLS159 are a PLS153 with four, six and eight flip-flops respectively, while a PLS179 is a PLS173 with eight registered outputs. Unlike PALs, though, FPLSs use either J-K or R-S flip-flops, and a typical, simplified, schematic is shown in Figure 1.26.

The advantage of J-K flip-flops can be seen if a simple state machine is examined. Consider a PIN detector, or a circuit for detecting if the number

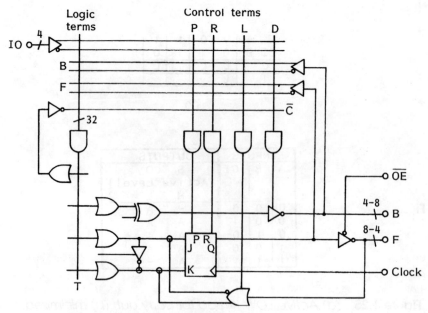

Figure 1.26 *FPLS simplified schematic*

Classic logic devices 27

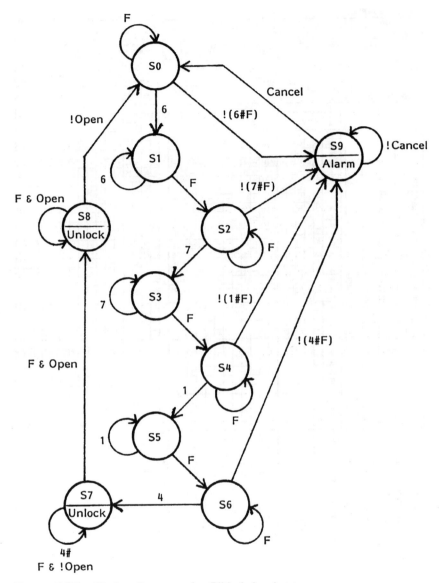

Figure 1.27 *State diagram for PIN detector*

entered from a keypad is valid, to unlock a secure door, for example. This can be incorporated into a PLS155, with the state diagram of Figure 1.27.

S0 is the start state, where the system waits for an input. Let us assume that the number to open the lock is 6714. Entering a '6' will send the state

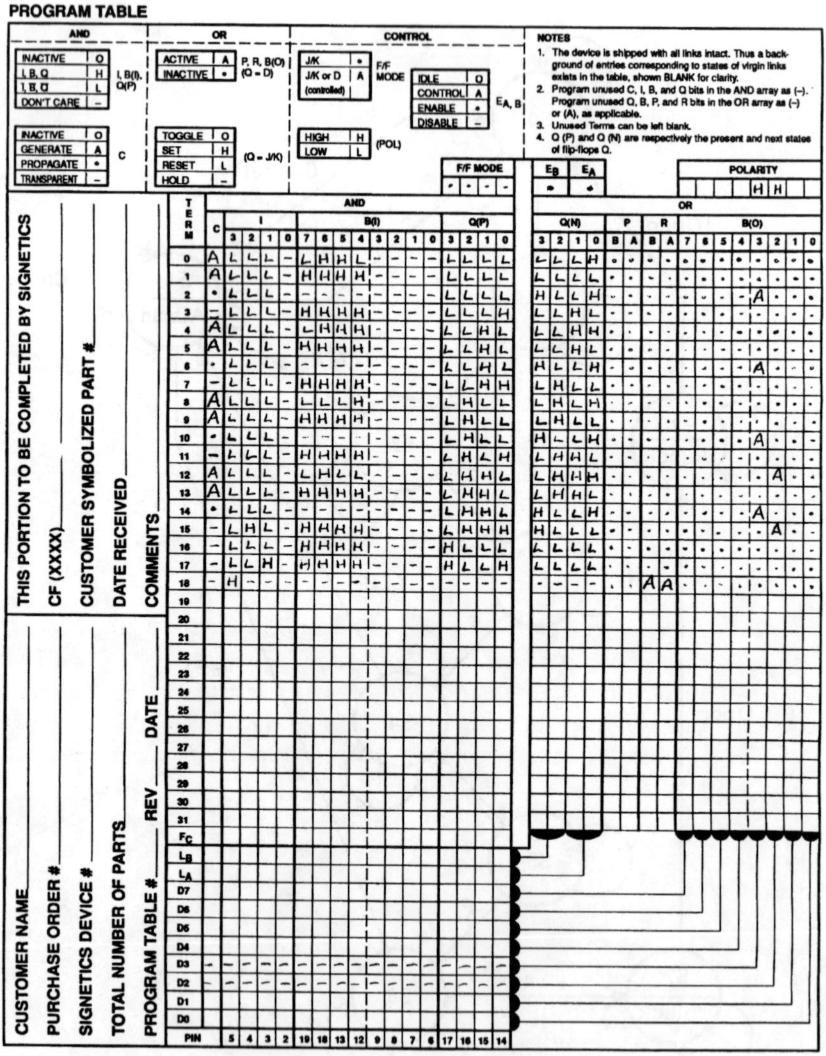

F/F MODE				E_B	E_A	POLARITY B(O) 7	6	5	4	3	2	1	0
•	•	–	–	•	•					H	H		

TERM	C	I 3	I 2	I 1	I 0	B(I) 7	B(I) 6	B(I) 5	B(I) 4	B(I) 3	B(I) 2	B(I) 1	B(I) 0	Q(P) 3	Q(P) 2	Q(P) 1	Q(P) 0
0	A	L	L	L	–	L	H	H	L	–	–	–	–	L	L	L	L
1	A	L	L	L	–	H	H	H	H	–	–	–	–	L	L	L	L
2	•	L	L	L	–	–	–	–	–	–	–	–	–	L	L	L	L
3	–	L	L	L	–	H	H	H	H	–	–	–	–	L	L	L	H
4	A	L	L	L	–	L	H	H	H	–	–	–	–	L	L	H	L
5	A	L	L	L	–	H	H	H	H	–	–	–	–	L	L	H	L
6	•	L	L	L	–	–	–	–	–	–	–	–	–	L	L	H	L
7	–	L	L	L	–	H	H	H	H	–	–	–	–	L	L	H	H
8	A	L	L	L	–	L	L	L	H	–	–	–	–	L	H	L	L
9	A	L	L	L	–	H	H	H	H	–	–	–	–	L	H	L	L
10	•	L	L	L	–	–	–	–	–	–	–	–	–	L	H	L	L
11	–	L	L	L	–	H	H	H	H	–	–	–	–	L	H	L	H
12	A	L	L	L	–	L	H	L	L	–	–	–	–	L	H	H	L
13	A	L	L	L	–	H	H	H	H	–	–	–	–	L	H	H	L
14	•	L	L	L	–	–	–	–	–	–	–	–	–	L	H	H	L
15	–	L	H	L	–	H	H	H	H	–	–	–	–	L	H	H	H
16	–	L	L	L	–	H	H	H	H	–	–	–	–	H	L	L	L
17	–	L	L	H	–	H	H	H	H	–	–	–	–	H	L	L	H
18	–	H	–	–	–	–	–	–	–	–	–	–	–	–	–	–	–
D3		–	–	–	–	–	–	–	–	–	–	–	–	–	–	–	–
D2		–	–	–	–	–	–	–	–	–	–	–	–	–	–	–	–
PIN		5	4	3	2	19	18	13	12	9	8	7	6	17	16	15	14

TERM	Q(N) 3	Q(N) 2	Q(N) 1	Q(N) 0	P B	P A	R B	R A	B(O) 7	B(O) 6	B(O) 5	B(O) 4	B(O) 3	B(O) 2	B(O) 1	B(O) 0
0	L	L	L	H	•	•	•	•	•	•	•	•	•	•	•	•
1	L	L	L	L	•	•	•	•	•	•	•	•	•	•	•	•
2	H	L	L	H	•	•	•	•	•	•	•	•	A	•	•	•
3	L	L	H	L	•	•	•	•	•	•	•	•	•	•	•	•
4	L	L	H	H	•	•	•	•	•	•	•	•	•	•	•	•
5	L	L	H	L	•	•	•	•	•	•	•	•	•	•	•	•
6	H	L	L	H	•	•	•	•	•	•	•	•	A	•	•	•
7	L	H	L	L	•	•	•	•	•	•	•	•	•	•	•	•
8	L	H	L	H	•	•	•	•	•	•	•	•	•	•	•	•
9	L	H	L	L	•	•	•	•	•	•	•	•	•	•	•	•
10	H	L	L	H	•	•	•	•	•	•	•	•	A	•	•	•
11	L	H	H	L	•	•	•	•	•	•	•	•	•	•	•	•
12	L	H	H	H	•	•	•	•	•	•	•	•	•	A	•	•
13	L	H	H	L	•	•	•	•	•	•	•	•	•	•	•	•
14	H	L	L	H	•	•	•	•	•	•	•	•	A	•	•	•
15	H	L	L	L	•	•	•	•	•	•	•	•	•	A	•	•
16	L	L	L	L	•	•	•	•	•	•	•	•	•	•	•	•
17	L	L	L	L	•	•	•	•	•	•	•	•	•	•	•	•
18	–	–	–	–	•	•	A	A	•	•	•	•	•	•	•	•

Figure 1.28 *State table for PIN detector*

machine to S1, where it will remain until the keypad is released and outputs an 'F' code, when it enters S2. Any input other than '6' or 'F' will cause a transition to S9 and set off an ALARM signal; this will continue until the CANCEL signal is set. A similar pattern is followed for the other three digits until S7 is entered, when UNLOCK is asserted. The system remains in S7 until the door is opened when we progress to S8. Finally, closing the door returns us to S0.

Each transition is entered into the state table of Figure 1.28 as an individual term. The signal names are as follows:

RESET	I3
CANCEL	I2
DOOR_OPEN	I1
DIGIT_INPUT	B7–B4
STATE BITS	Q–Q0
ALARM	B3
UNLOCK	B2

Thus, term 0 defines the jump from S0 to S1, and term 1 the hold condition for S0. The latter is not normally necessary because J-K flip-flops hold when both inputs are LOW but, if both terms are fed to the complement term (the 'A' in the 'C' column) term 2 uses the complement term to trigger the jump to S9 when both terms 0 and 1 are false while the state machine is in S0. Similarly, term 6 triggers a jump from S2 to S9 when terms 4 and 5 are false and the system is in S2. The start state is entered by the RESET signal being HIGH.

The state table, as drawn, has not been minimized at all but this can be done by inspection in many cases. For example, terms 2, 6, 10 and 14 can be combined into a single term. On the other hand, it could be that the designer might prefer not to minimize as some clarity can be lost when this is done, unless it is necessary to make enough transition terms free to fit the logic into the FPLS.

As this example shows, a relatively complex system can be accommodated in a 24 to 28 pin PLD and can perform a useful function as a microprocessor peripheral. This complexity is close to the upper limit of what can be achieved in this size package so we next need to look at the options in larger devices. More pins often means more complex logic, either because there are more signals to cope with, or because several functions can be combined into a single package. As logic complexity increases, more sophisticated methods need to be used to design the logic; in the rest of the book we will look at the different structures used to accommodate higher levels of logic, and the methods used to design with them.

2 Programmable LSI techniques

2.1 LSI PAL devices
2.1.1 Overall structure

One approach to making PLDs with a higher logic content is to integrate several small PALs in one package. Different manufacturers may differ in detail in the way they do this but they all have certain features in common. The basic structure of an LSI PAL-type device, or CPLD (complex PLD), is shown in Figure 2.1.

There are four blocks to consider in each device, the input cells, an interconnection matrix, the logic blocks and the I/O cells. We can discuss these in detail for each device family, but examine the basic attributes of each here.

Input cells and I/O cells may be taken together as some families have no, or only a few, separate inputs and use I/O cells for this function. In general, an I/O cell connects the logic blocks to the outside world and features a TTL or CMOS interface. Most devices operate with 0.4 V and 2.4 V output logic levels and therefore need some conversion if they are to drive true CMOS logic; they have high impedance inputs and can be driven by CMOS without any problem.

The outputs always have a tri-state capability which can be permanently enabled or disabled, or controlled by internal logic. They can function as dedicated inputs, dedicated outputs with optional feedback or as bussable outputs. The advantage of being able to choose the function of all I/Os is that the input and output sites are not pre-determined and this gives extra flexibility to the PCB layout.

In small PLDs, every input to the device may usually be connected to every product term. Even a quite large PAL, such as the 22V10, will have a manageable number of fuses with this universal connectivity. The 22V10 has 132 product terms with 44 potential connections to each,

Programmable LSI techniques 31

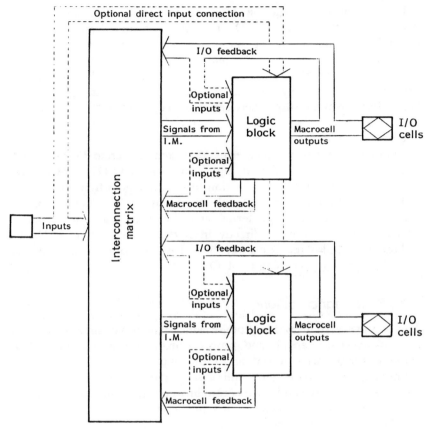

Figure 2.1 CPLD block diagram

giving a total of 5808 fuses not counting configuration sites. A PLD in a PLCC84 package may have 72 I/Os; if each I/O has an average of just four product terms there will be 4 × 72 × 144, or 41 472, programmable cells. This does not allow for any buried logic cells. If each I/O has a buried macrocell associated with it the total number of fuses is quadrupled.

While such a total is not outside the possibilities of technology there are performance penalties with this approach, as we shall see in the next section, and the more fuses there are the more expensive the chip and the longer it takes to program. The solution is an interconnection matrix which selects a reduced number of inputs to feed each logic block. It is best, therefore, to consider each logic block as an individual PAL within the larger PLD, each PAL being connected to the others via the interconnection matrix.

Looking at each logic block as an individual PAL simplifies the description of the architecture. Discrete PALs contain at least eight, and up to sixteen product terms per output but, very often, only two or three terms are actually used in many applications. This variability gives the PLD manufacturer a problem; how many product terms does he provide with each macrocell. In many cases the answer is to incorporate some method where terms may be shared between macrocells within each logic block. Then, if a function does not need all its available terms they can be used by one of its neighbours.

Another area where product terms are wasted in discrete PALs is the I/O macrocell used as a dedicated input. In many CPLDs there are two feedback paths to the interconnection matrix; one goes from the I/O pin and the other from the macrocell output. In this case, if an I/O pin is used as a dedicated input the logic terms can still be used as a buried macrocell which cannot be accessed directly from the outside. Some PLDs also include macrocells which are always buried as a way of increasing the logic capability without increasing the I/O count.

2.1.2 Performance criteria

The two principal measures of performance which will affect selection of a logic device are **speed** and **power consumption**, but the balance between them varies according to the application. Clearly, in a fast computer, speed is the most important factor whereas, in a portable telephone, power would be the controlling consideration. In many cases a compromise can be achieved, as 'it must be faster than 15 ns but the less power it takes the cheaper we can make the power supply'.

Propagation delay, the most common measure of speed, is almost entirely determined by the time taken to charge and discharge the capacitance associated with the components and interconnections used to build the device. Once again, we can split a PLD into sections to see which areas are critical and reliant on the design of the chip rather than just processing.

Starting with the input, this is probably the simplest part of the PLD as its function is just to buffer external logic levels to the internal levels. Delay time will be largely process-dependent as the input buffer will be designed to the minimum geometry allowed by the process, there being no heavy drive needed to interface to the internal logic. Some devices do feature input latches or registers which will be optional and, therefore, add no delay when by-passed. When they are being used it is, again, most likely that they will be a minimum geometry design and their time penalty be entirely process-dependent.

Outputs determine the drive available for interfacing to external circuitry and, in this way, are no different from any other logic device. The drive capability of an output stage is usually characterised by the output short-circuit current.

The interconnect matrix is one section of a CPLD whose performance can be affected by the way in which the chip is designed. The matrix is an array of switches with as many inputs as there are device inputs, I/Os and macrocell feedbacks, and there are outputs to each of the logic blocks. Each switch is like a small capacitor hanging on the input line, whether it is connected to a logic block, or not, so the number of inputs to all the blocks determines the capacitance of the interconnect matrix.

The time taken to charge and discharge a capacitor is determined by the current which is used for this, so one way to speed up the matrix would be to increase the current driving each of the block inputs. This would also have the effect of increasing the power consumption of the whole chip. There are two consequences of doing this; firstly it increases the size, and hence cost, of the power supplies in the overall system. Secondly it increases the heat produced by the chip and, therefore, its running temperature, and will also have implications in the cooling needed for the system in which it is being used.

An alternative approach is to limit the number of signals which are switched into each of the logic blocks. This will reduce the total capacitance loading each AND term and hence the delay time for a given level of current drive. It also means that some of the input and feedback signals are not available for use by some of the AND terms.

The logic block itself offers only a little scope for performance optimization. We have discussed the effect of changing the size and drive level of the input product terms; limiting the number of terms in each block will save power, or allow more power to be used for each term, but this also limits the complexity of the logic function which can be implemented. At the heart of the block is, usually, a flip-flop with surrounding switches to route the logic signal, as in the 22V10.

Many LSI PAL devices also include product term switching which can allocate some or all of the product terms to adjacent logic blocks. This reduces the waste when blocks are only partially used, or used as direct inputs.

Although the product terms consume power in normal operation some devices contain a 'turbo' switch. In non-turbo mode the logic array is only powered when input signals are changing, the output signals being latched. An activity detector switches on the array power and enables the output latches when a change is detected at any input. Because this causes extra delay through the device, setting turbo on keeps power applied permanently giving high speed at the expense of high dissipation. This circuit configuration is shown, diagrammatically, in Figure 2.2.

The features which characterize CPLDs, then, are logic blocks with wide input gating, a central switch matrix, one-to-one correspondence

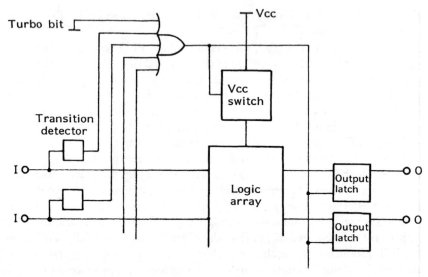

Figure 2.2 *'Turbo switch' circuit*

between logic blocks and I/O pins and a compromise between speed, power and internal connectivity.

2.2 Logic cell arrays

2.2.1 Typical architecture

Although the term 'logic cell array' is strictly applicable to one manufacturer's product, it does describe the second generic programmable LSI structure quite accurately. The floor plan of a typical LCA is shown in Figure 2.3, and contains three basic elements, I/O cells, logic cells and routing channels. Most devices in this class use RAM cells for configuration and signal routing.

Signals enter and leave the device through I/O cells which have connections to the busses in the routing channels. The logic cells also have programmable connections to the routing lines which run past them and, in this way, the signals are propagated round the chip. The logic blocks themselves are usually more complex than a typical PAL-type macrocell, but with far fewer inputs, typically only four or five. Thus, the logic structures can be fabricated from pure CMOS.

Unlike CPLDs, whose I/O pins are usually associated with a particular logic block, LCA I/Os are independent of the internal logic. Any pin can be an input or an output and all the I/O cells are identical, although they may share their function with the signals needed to set the device up in the

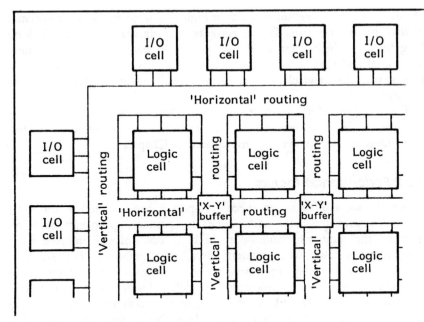

Figure 2.3 LCA floor plan

configuration phase. Complexity of the I/O cell varies according to device family, but they will normally have the usual functions such as tri-state capability on outputs and the possibility of latching or registering input signals.

As stated above, the logic cells themselves tend to have the possibility of implementing quite complex functions but with no more than six inputs. Implementing wide gating functions may require cascading over two or more levels of logic. A CPLD can usually cope with a width of twenty or more signals in a single level of gating. An LCA design will, therefore, look more like a discrete logic design than a PAL, and logic minimization may be performed more profitably with this in mind.

The routing channels in an LCA owe more to masked ASICs than PALs also. Most LCAs include different lengths of routing within a device for, as we shall see, many of the factors which determine device performance are connected with routing. Short tracks are usually available for routing signals between adjacent cells while longer tracks may be needed for carrying signals across larger areas of the chip. One or more additional tracks may be provided for clocks and other global signals, which may need a high fan-out.

The longer a routing track the higher its capacitance so performance is optimized by keeping track lengths as low as possible, which is helped by placing cells close together when their logic functions are closely associated in the circuit schematic. Often, **repeaters** or **drivers** are inserted along long tracks to help improve the delay due to track capacitance. Where horizontal and vertical routing channels cross they may be connected by **programmable switches** in order to divert signals to other rows or columns.

The way in which each manufacturer has attacked the problem of providing sufficient routing resources without over-burdening the chip with interconnections will be investigated in Chapter 5.

2.2.2 Performance factors

As before, we will focus on speed and power consumption as the most important measures of performance. In practice, propagation delays between the different elements in an LCA can only be inferred from measurements made with different circuit configurations. For example, the difference in delay time between a single-level logic circuit and a two-level logic circuit should give the delay through a logic cell. The problem in making a judgement between two different circuits lies in other factors, such as the interconnect delay. Performance will depend as much on the way in which circuits are laid out as on the delay through circuit elements.

Furthermore, different arrays have logic cells of different complexity, so performance can also depend on the appropriateness of the logic function and the logic capability of the target device. At a simple level, a six-input gate may need two stages of gating in one array but only one in another, making the second appear faster irrespective of the intrinsic delay of the two devices. One good measure of the true speed of the internal logic is the maximum frequency at which a flip-flop will operate. This can be gauged by configuring a cell as a divider and finding the frequency at which the output disappears.

Detailed methods of measuring device performance for both standard configurations and custom circuits will be discussed in a later chapter.

Because the internal logic cells in most LCA devices use relatively narrow gating, true CMOS circuitry is used in building the internal logic cells. The stand-by power is therefore comparable, in most cases, to that of the same circuit built in discrete CMOS logic. The only factor which usually causes a problem is any level shifting needed to interface the inputs to TTL levels.

As with discrete CMOS logic, power consumption starts to rise as soon as signals in the circuit start to change, and internal and external nodes need charging and discharging. Because the capacitance involved with internal connections is significantly lower than that of external connections, an LCA will usually dissipate less power during clocked operation than the equivalent discrete circuit. The actual power will depend on the number of nodes which are switching at any one time and will need some complex analysis to make an accurate prediction of power. It is often possible to calculate the predicted power consumption from within the design environment.

2.2.3. Device configuration

The devices in this class of FPGA usually hold their configuration data in a shadow RAM. Every interconnection site and every logic configuration bit is defined by one bit of this RAM. Because static RAM is a volatile storage medium, on power-up an LCA will not be configured at all. The configuration data must be stored in some non-volatile device, such as an EPROM, or it may be booted from disk in the same way as a computer operating system.

Some devices can use standard components, such as 27Cxxx EPROMs, while others are tailored to a specific memory; detailed data on each type will be presented in Chapter 5. A typical configuration set-up is shown in Figure 2.4.

There are two issues raised by the use of external memory components to configure LCAs. Firstly there is the problem of security; one of the attractions of programmable logic devices is the ability of most of them to read-protect the data after programming, making it difficult to copy the function programmed into the PLD. If the programming data has to be read in each time the system is powered up, this principle is defeated at the outset. Even if the data in the external store is protected somehow, a logic analyser will soon decipher the bit stream which is used to program the LCA.

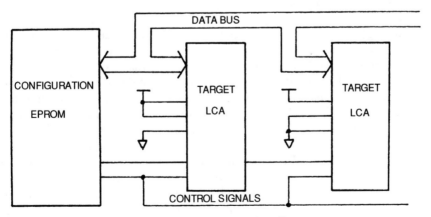

Figure 2.4 *Typical LCA configuration circuit*

A less contentious issue is the use of two devices to perform a logic function which, in other families, is contained within a single device. This may be overcome by combining the data for LCA configuration with some other data, such as the operating program for a microprocessor, but it still leads to complications in the hardware design for the overall system. If several LCAs are used in one system they can share a single EPROM, in most cases, which reduces the hardware overhead to a fraction of a device.

The upside of this arrangement is that reconfiguration of a system can be managed by software even, in many cases, while the system is actually operating. This may be a simple change, such as redefining a memory map to re-allocate resources, or a partial or complete change of logic function. In Atmel devices, part of the device can be reconfigured while the rest is still functioning; the name **Cache Logic** has been coined to describe this function, by association with **cache memory**, which buffers and speeds up data exchanges in computer systems.

2.3 Antifuse FPGAs

2.3.1 Antifuse technology and architecture

The architecture of antifuse FPGAs is practically identical to LCAs, that is, uncommitted I/O cells surrounding logic cells and routing channels. Any differences are due mainly to the two different technologies involved. The main difference is in the resistance of the transmission elements. LCAs use an MOS transistor, which has a relatively high on-resistance compared with an antifuse, once it has been blown.

Whenever a signal passes through a fuse element it must charge up the capacitance of the component to which it is connected. The capacitance of tracks and gate inputs will not differ significantly between the two technologies so the series resistance of the fuse makes a significant difference to device performance. Antifuse devices can, therefore, use an architecture which has longer connecting tracks than LCAs.

In practice they are often supplied with tracks of differing lengths within a single routing channel. The actual track length required for any specific connection can be optimized by choosing an appropriate path in the routing channel. By minimizing interconnect capacitance and series resistance, in this way, the interconnect delays are minimized and there is no need for repeaters or drivers within the connection structure.

Another trade-off is the size of the logic cell and the amount of interconnection. In a practical device it must be possible to connect the logic cells together even when approaching 100% utilization. One way to improve connectivity is to make the logic modules quite complex; by incorporating more logic in each module, less connections between modules are needed. This is the trend in LCAs.

If track delays are less significant the logic modules may be made simpler, and complex functions built by using short lengths of track to connect them. Module usage is likely to be improved in this way, with less wasted logic functions and silicon area. Antifuse FPGAs are, therefore, able to use a smaller grained structure than LCAs and, usually, offer a less wasteful solution to most logic circuits.

Although this approach may need more interconnects, the antifuse technology helps to reduce the cost impact of the on-chip area. Antifuses occupy about one micrometre square and fit inside the aluminium tracks, thereby taking up no extra space in the array. RAM-based FPGAs need a RAM cell and MOS transistor at each crossing point, and these add to the area occupied by the interconnect.

Granularity is an important feature of ASICs as it determines the efficiency with which logic functions can be implemented. Masked ASICs, for example, are usually fine grained as they have low resistance vias to connect logic blocks to the tracks. A logic cell is just a small gate and more complex functions are easily built without incurring much speed penalty from the interconnect. Antifuse FPGAs have a more complex basic cell; a flip-flop with some universal gating is quite common. A typical LCA logic block has two flip-flops driven by some fairly complex combinatorial logic.

By way of example, a 4000-gate LCA (Xilinx XC4004A) has 144 logic blocks while a 4000-gate antifuse FPGA (Actel A1240A) has 684 modules. In order to achieve 4000 routable gates, a sea-of-gates gate array may need 7000 gates in its basic array, where one gate in the array corresponds to a single cell but also has to act as part of the routing resource.

2.3.2 Device performance

Most of the remarks about LCA devices also apply to antifuse FPGAs. The logic cells are true CMOS and consume 'zero power' at zero frequency. Logic circuits are not very useful at zero frequency and as soon as signals begin to switch, power is dissipated by the charging currents for the parasitic capacitances. In practice, this current is negligible below 1 MHz but becomes significant above this frequency.

Within an FPGA there are two sources of power dissipation, the active circuit elements and the interconnects. For a given feature size, capacitance of circuit elements will be practically the same, irrespective of architecture. Any difference in consumption will be due largely to the interconnect capacitance. We have seen that antifuse FPGAs tend to use more tracks than LCAs, as the LCA logic cells can perform more complex functions. On the other hand, LCAs may use longer tracks and higher capacitance connections between tracks, so the comparison may well depend on the actual logic being implemented.

The same arguments will also apply to dynamic performance, with the added factor of the resistance of the two types of programmable connection. Logic requiring the connection of large numbers of gates will usually perform better in antifuse FPGAs, because of the superior performance of the antifuse. Where logic functions can be concentrated into the more complex LCA cells, antifuse devices will probably use more interconnects and show a poorer dynamic performance. If we extend the argument to include CPLD structures, the trend is more pronounced. They will perform well in highly structured circuits, where the macrocells can implement a whole logic function, but deteriorate badly as soon as repetitive feedback has to be introduced in more fragmented circuits.

2.3.3 Using antifuse FPGAs

The main drawback with antifuse FPGAs centres on the programming situation. Even a small circuit can take three minutes to programme while the larger devices may take ten minutes. This compares with a few seconds for most CPLDs structures and no time at all for LCAs, which are programmed in-circuit each time they are powered up. A single programming station, with four sockets, is limited to between 200 and 600 circuits a day and, therefore, adds a significant overhead to the basic cost of the device.

The upside to this limitation on production throughput and increased cost is a double benefit. A circuit using an FPGA will start working the instant it is switched on, and requires no overhead in program storage or extra design work in arranging for the device configuration function. The other advantage is security. It is difficult to read back some FPGAs, but they also contain security fuses to prevent this from being done. CPLDs also contain security cells to prevent direct reading, although sophisticated analysis may defeat device security. LCAs, as we have seen, are much more difficult to keep secret.

3 Designing FPGAs

3.1 Introduction

The one thing that all the devices described in this book have in common is that they are all logic devices. Because of this, they all share a common approach to the design process. In principle, any logic circuit will fit into any FPGA within the restraints of logic content and connectivity. Crudely, as long as there are enough gates and enough I/Os the choice of FPGA does not affect the way in which the logic is defined. In practice this is not quite true, but we shall look at dedicated approaches to the various families at a later stage, and cover the general points of design in this chapter.

There are only two general ways to design logic; the required function may be described in terms of some written language, or it may be drawn in some symbolic manner. Written methods include logic equations, state equations and hardware description languages; symbolic descriptions are covered by circuit diagrams and state diagrams.

All design packages make use of one or more of these categories of input to define the logic. Once the logic has been defined it is usually not dependent on any one target device or architecture, so this may be looked on as just the first stage in a design. The second stage is to ensure that the defined logic does the job which it is intended to do. This is achieved by simulating the design; that is, applying inputs to a software model of the design and checking that the outputs are as expected.

At this stage the target device can be considered. A translation from the general logic definition to specific architectural units is undertaken; abstract logic is mapped onto the physical components of the target device and potential internal connections decided. This third stage, device fitting, shows whether the target device has the capability to contain all the logic functions in the abstract design, but does not guarantee that the programmed device will work in the designers circuit. This can be taken a step closer by post-layout simulation.

Now that a real device is involved, with internal components and connections of predictable performance, the time-dependent factors can be added to the simulation result. Not only can we confirm that output Y = input A AND input B, but we can also predict that signal Y will go HIGH within 5 ns of both inputs A and B going HIGH. Thus we can program a device and plug it into our circuit with a high confidence of success, provided that we have covered all critical eventualities in our simulation.

We can now examine each design method in detail and see how they measure up to the requirements outlined above.

3.2 Logic equations

3.2.1 Input methods

Logic equations have been the standard method for designing PLDs ever since their introduction to the market in the late 1970s. As we saw in Chapter 1, the early PALs were simple sum-of-product devices whose outputs depended on a straightforward logic relationship between the inputs, often without any feedback, or other complications. Logic equations are the most usual way of defining the logic content of classical PLDs.

Compilers for logic equations commonly consist of four sections, an introduction where the drawing number, designer's name and company and a brief functional description plus other relevant information can be listed, a pin-out definition, an equations section and a simulation segment. The introduction needs no further discussion as it merely annotates the design with information needed for future reference. The pin-out definition section is also self-explanatory. It lists the signals used in the design and allocates them to the device I/O pins.

We can illustrate the logic equation part of the design input by looking at some simple examples of the use of equations. The most common application for PALs was as address decoders in microprocessor circuits, where standard TTL parts did not have the input width or flexibility to provide an economic solution.

For example, an equation such as:

!Y = A15 & A14 & !A13 & !A12 & A11 & !A10 & !A9 & A8

was, and still is, commonplace in PAL design sheets. It fits comfortably into one eighth of a standard PAL whereas it would need the best part of two gate packages to implement it in discrete logic. It is also easy to understand what is meant by this equation; when the processor puts out address 36xx,

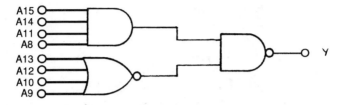

Figure 3.1 *Discrete logic address decoder*

this output is taken LOW and will, presumably, enable some peripheral chip. Figure 3.1 shows how this circuit may be implemented in discrete logic and, although not completely incomprehensible, it is not as immediately apparent as the logic equation. If there were six or seven decoded outputs, all looking like Figure 3.1, there might well be some confusion, especially without some annotation on the drawing. Annotating equations is straightforward enough; our address decode may be commented as:

!Y = A15 & A14 & !A13 & !A12 & A11 & !A10 & !A9 & A8;
    decode of 36XX

the semi-colon delimiting the comment from the actual equation.

More complex functions may be defined as logic equations, with equal clarity, by using various shorthand techniques. For example, a four-bit identity comparator, which is constructed from four exclusive-OR gates and an AND gate may be defined in the following way:

EQ0 = A0 & B0 # !A0 & !B0; 'zero' bits equal
EQ1 = A1 & B1 # !A1 & !B1; 'one' bits equal
EQ2 = A2 & B2 # !A2 & !B2; 'two' bits equal
EQ3 = A3 & B3 # !A3 & !B3; 'three' bits equal
EQ = EQ0 & EQ1 & EQ2 & EQ3; all bits equal

Using the exclusive-OR symbol ('#') instead of the expanded logic definition would make the equations even clearer, and would be understood by most logic compilers.

We can also define a state machine with logic equations, as we can show with the system defined in Section 1.3.3. This was a four-digit PIN detector whose state diagram is reproduced in Figure 3.2; the application, as described previously, is a keypad door lock. It is unlocked by entering the code '6714'; any other code sets off an alarm which can be cancelled with a remote reset. The machine is also reset when it detects that the door has been closed after a successful entry.

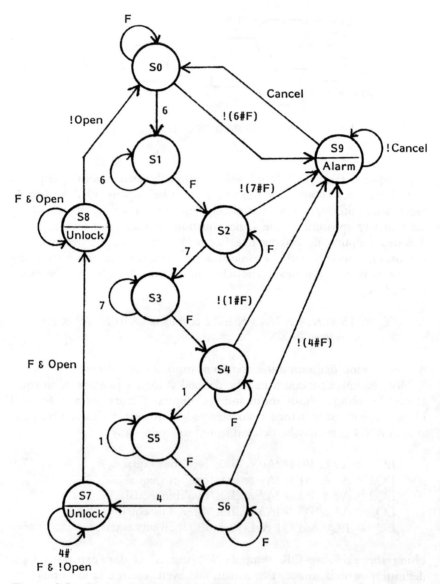

Figure 3.2 State diagram for 'PIN' detector

Physically, the state register has four bits, outputs Q3–Q0, with four inputs I3–I0 for the code entry, DOOR for the door close signal and ALARM for the alarm cancel signal. The keypad decoder sends out a digit encoded by I3–I0 or 'F' is no key is being depressed. System outputs are UNLOCK to enable the door, and SOUND to set the alarm.

First, we have to define the states and the input codes, as follows:

```
S0  = !Q3 & !Q2 & !Q1 & !Q0
S1  = !Q3 & !Q2 & !Q1 & Q0
S2  = !Q3 & !Q2 & Q1 & !Q0
and so on through
S9  = Q3 & !Q2 & !Q1 & Q0
INF = I3 & I2 & I1 & I0
IN7 = !I3 & I2 & I1 & I0
IN6 = !I3 & I2 & I1 & !I0
IN4 = !I3 & I2 & !I1 & !I0
IN1 = !I3 & !I2 & !I1 & I0
```

If we are designing for a device with D-type flip-flops, the state transitions must be defined by noting which bits have to be set HIGH for each state transition, as follows:

```
Q3.D = S7 & INF & !DOOR; transition S7 to S8
 #  S8 & !DOOR; hold in S8
 #  S0 & !(IN6 # INF); transition S0 to S9
 #  S2 & !(IN7 # INF); transition S2 to S9
 #  S4 & !(IN1 # INF); transition S4 to S9
 #  S6 & !(IN4 # INF); transition S6 to S
 #  S9 & !ALARM; hold in S9
Q2.D = S3 & INF; transition S3 to S4
 #  S4 & INF; hold in S4
 #  S4 & IN1; transition S4 to S5, etc.
```

The outputs are simply defined by:

```
UNLOCK = S7;
SOUND  = S9;
```

The notation 'Q3.D' implies that this function is applied to the D-input of the flip-flop driving the Q3 output. At the active clock edge, the output will be set HIGH if the function is TRUE.

From the first few lines it is apparent that equations do not give a transparent view of the function being implemented. However, by considering each transition it is possible to generate a set of equations which, when mapped onto the device, will produce a working part. The logic compiler will minimize each equation and, if there are enough product terms driving each flip-flop, the mapping will be successful.

We shall see, below, how state machines can be specified in a way which is directly associated with their function.

3.2.2 Simulation

The final segment of a logic design is usually a simulation. This has two functions; it checks that the logic will operate as intended, and it produces a set of test vectors for performing functional tests on the device after it has been programmed. Simulation can be defined by means of a text-based entry, or with a truth table, in most logic compilers.

We can show how the door lock can be simulated with an example of each method. The PALASM format would yield the following:

```
SIMULATION ;keyword for the simulation segment
TRACE_ON I3 I2 I1 I0 DOOR ALARM Q3 Q2 Q1 Q0 RESET CLK ;defining
     the signals we wish to view on the simulation output. Reset
     is added to initialize the state machine
SETF I3 I2 I1 I0 /DOOR /ALARM RESET ;initial values of the inputs
     at reset
SETF /RESET ;release the reset condition
CLOCKF CLK ;clock the state machine
CHECK /Q3 /Q2 /Q1 /Q0 ; check that it is state '0' (an
     alternative notation is - CHECK S0)
SETF /I3 I2 I1 /I0 ;put '6' onto the inputs
CLOCKF CLK ;clock it again
CHECK S1
CLOCKF CLK
CHECK S1 ;check that it holds in S1
.
.
.
TRACE_OFF ;end trace at finish of simulation
```

The same test sequence in truth table format would appear as in Table 3.1.

Although the truth table requires more typing effort, the result is clearer in terms of the device operation and the output is automatically checked on each line, not just when specified as in the text-based method.

Table 3.1

[I3,	I2,	I1,	I0,	DOOR,	ALARM,	RESET,	CLK] →	[Q3,	Q2,	Q1,	Q0]
[1,	1,	1,	1,	0,	0,	1,	0] →	[0,	0,	0,	0]
[1,	1,	1,	1,	0,	0,	0,	C] →	[0,	0,	0,	0]
[0,	1,	1,	0,	0,	0,	0,	C] →	[0,	0,	0,	1]
[0,	1,	1,	0,	0,	0,	0,	C] →	[0,	0,	0,	1] |//
and so on.

Simulation results may usually be viewed in tabular form or as a waveform display after compilation of the logic equations. Any discrepancies between the expected result of the simulation and the actual result will show up and allow modification of the logic equations to give the desired logic function. Compilation of logic equations, together with the simulation segment will give a programming file with test vectors where this option is possible. Most PAL-type devices will accept programming files with vector testing; most FPGAs are either configured in-circuit or programmed on dedicated programmers which do not allow for vector testing. In-circuit testing, using JTAG protocols, is more usual for FPGAs and will be covered in a later chapter.

3.2.3 Logic equation shortcomings

There are two problems with using logic equations in FPGA designs. Firstly there is no global standard for the symbols and syntax; in this book we have standardized on the ABEL symbols, but other standards, such as PALASM are equally valid. This is not an obstacle in itself for these different standards have worked very well for classical PLD designs, but it would be very useful to have a universal standard which would be accepted by any FPGA design system.

More important is the scale of FPGAs compared with PLDs. A fairly complex PLD such as the 22V10, contains over one hundred product terms, each of which can be defined by some equation like our decode example above. To try to design with equations, even at this scale, can become nearly incomprehensible. Writing down the equations is a time-consuming task in itself, but trying to decide where modifications should be made in the event that simulation throws up a mistake, for example, may prove even more arduous.

The architecture of FPGAs dictates that logic equations are not the best way to define their logic content. CPLDs have a fixed structure which means that a logic equation maps directly into the AND-array of the logic cell. The only variable is the way in which signals are routed from the I/Os into the logic blocks, but even this task can be approached in a fairly mechanistic way, for there are fixed paths for the signals to travel inside the device.

In an FPGA the structure is much freer. For a start, there is not a fixed two-level AND-OR structure for a one-to-one association with the typical sum of products logic equation. This may have to be broken down into a chain of gates in order to achieve the required input connectivity. If the same logic expression is used in two different equations the inference is that it is recreated at each location in the FPGA where the overall equations are implemented. It may be more efficient to generate it only once and route it to the second area of the FPGA. This will save logic modules but use routing resources.

We have stressed that FPGA structures are more like an integrated form of discrete logic, and it is most unusual to design PCBs for discrete logic with logic equations. While logic equations are useful and help visibility for PAL-type PLDs, they are not the ideal way to design FPGAs.

3.3 State equations

3.3.1. State machine basics

A state machine is a system which, as its name implies, can exist in a number of stable states. Each state is usually defined by a unique number stored in a set of flip-flops called the state register. Inputs to a state machine are those signals which can influence the sequence in which the states are entered, and a clock to define the time intervals at which the inputs are sampled and the decision made as to whether the state register is changed, and which state should be entered next. The outputs from a state machine may depend on the state register only, in which case it is called a Moore Machine, or they may be a logical combination of inputs and state register, when it is known as a Mealy Machine.

A state machine may be described by a state diagram; our door lock example was shown in Figure 3.2. The ten possible states are each represented by a circle labelled with the value of the state register for that state. Transitions between states are represented by arrows labelled with the logic condition which enables that transition. The 'main sequence' runs vertically downwards from S0 to S8, wrapping around back to S0 when the sequence ends with the door closing again.

Transitions to S9 are triggered by incorrect key depressions. Each state has a hold condition which is shown by an arrow wrapped round back to the same state. Thus, keying '6' triggers a jump from S0 to S1 but, if key '6' remains depressed for more clock cycles the state machine remains in S1. Releasing the key sends 'F' to the inputs and triggers the jump to S2, ready for the next key push.

The outputs are straight decodes from the states so we have designed a Moore machine in this case.

3.3.2. State equation syntax

Most PLD and FPGA logic entry systems allow logic equations and state equations to be mixed in the same design file. The syntax for entering state equations varies from system to system but the commonest methods use IF-THEN-ELSE statements or CASE statements.

In either case it is mandatory to define the states in terms of the state register elements. This may be done with the following syntax:

```
[Q3,Q2,Q1,Q0]
S0 = 0000b;
S1 = 0001b;
S2 = 0010b;
```

and so on, or by:

```
STATE
S0 = !Q3 & !Q2 & !Q1 & !Q0;
S1 = !Q3 & !Q2 & !Q1 & Q0;
S2 = !Q3 & !Q2 & Q1 & !Q0;
etc.
```

The IF-THEN-ELSE structure may be written as:

```
WHILE [S0]
    IF I3 & I2 & I1 & I0 THEN [S0]
    IF !I3 & I2 & I1 & !I0 THEN [S1]
    ELSE [S9] WITH SOUND
WHILE [S1]
    IF !I3 & I2 & I1 & !I0 THEN [S1]
    IF I3 & I2 & I1 & I0 THEN [S2]
        .
        .
        .
WHILE [S9]
    IF ALARM THEN [S0]
    ELSE [S9] WITH SOUND
```

The ELSE statements, above, have different effects. In the [S0] statement, the ELSE sends the machine to state [S9] if the input is not 'F' or '6'; in the [S9] statement it defines the HOLD condition.

The WITH operator defines a combinatorial output. In a Moore machine, as in this case, an output will always be associated with the same state; in Mealy machines, output conditions may depend on the path by which a state is reached.

The CASE construct defines conditions to be tested, and the action to be taken when the condition is TRUE. In our door lock example, we could define the state machine with the following CASE statement:

```
CASE (ALARM, DOOR, Q[3..0])
BEGIN
    #h00:       CASE (I[3..0])
        BEGIN
        #hF:    BEGIN Q[3..0] = #h0 END
```

```
        #h6:        BEGIN Q[3..0] = #h1 END
        OTHERWISE:     Q[3..0] = #h9 END
        END
   #h01:         CASE (I[3..0])
        BEGIN
        #hF:        BEGIN Q[3..0] = #h2 END
        #h6:        BEGIN Q[3..0] = #h1 END
        END


   #h09:         BEGIN Q[3..0] = #h9 END
   #h29:         BEGIN Q[3..0] = #h0 END
   OTHERWISE:       Q[3..0] = #h0 END
   END
```

This illustrates the use of nested CASE statements. It is necessary, in this example, to allow the default jumps to [S9]; if the input conditions were not nested within each present state CASE, every possible input combination would have to be specified to define the jump to the error state. By nesting, the OTHERWISE operator takes care of defaults as the ELSE operator does in the IF-THEN-ELSE construct.

3.4 Hardware description languages

3.4.1. HDL derivation and overview

HDLs (hardware description languages) are a more generalized method of describing the behaviour of logic systems than logic equations, although they do embody some of the characteristics of logic and state equations. In this section we will concentrate on one particular language, VHDL, which is becoming an industry standard spurred on by MIL STD 454L, which requires all ASIC designs for the USA Defense Department to be documented in this language.

It was devised as part of the VHSIC (very high speed integrated circuit) project to allow complex ASICs to be specified and simulated without reference to any specific technology. Having specified a circuit in this way, it should be transferable to any process or manufacturer with guaranteed reproducibility.

VHDL (VHSIC HDL) describes logic systems from a top-down architectural standpoint. A system is visualized as a set of 'black boxes', called entities, with a set of interfaces. The top level entities may be broken into successively less complex functions until the bottom level is reached; this may be a gate-level description of the function.

Because each level of the logic hierarchy is specified uniquely, each may be simulated to check both syntax and logic function. The lowest level may be validated first so that a completely tested system is built from the bottom up. This is followed by a synthesis step which translates the whole design to the logic cell level after which it is simulated at gate level with built-in timing parameters.

This hierarchical design allows whole systems to be defined without specifying technology, or even partitioning into devices. The whole process is akin to designing software in a high-level language, with the low-level entities playing the same role as subroutines. A complete system may be defined and then tested without specifying a target device. The modules are designed separately and may be stored and used over again in future designs; in effect a library of functions is generated for re-use in new designs.

3.4.2. VHDL logic specification

Examples of VHDL definitions can show the difference between basic equations and HDL constructs. A four-bit adder may be defined as follows:

```
entity ADDER4 is
    port( A,B : in INTEGER range 0 to 15;
        C : out INTEGER range 0 to 15 );
end ADDER4;
```

The entity section defines the signal interfaces and sizes. Defining a range of 0 to 15 implies that each signal has four bits. It must be followed by an architecture section to define the logic relationship between the signals. This may be written as:

```
architecture BEHAVIORAL of ADDER4 is
begin
    C <= A + B;
end BEHAVIORAL;
```

The type of architecture, in this case BEHAVIORAL, is arbitrary as it is ignored in the compilation process; it is good practice to make it relevant to the way in which the architecture is being defined. Here, we are using a high-level description which will be understood during simulation but might need synthesizing at a lower level to achieve a good performance.

Lower level definition is generally specified as RTL (Register Transfer Level) which does not need to involve registers. The '<=' symbol represents the function 'signal assignment'; in the adder, the value of A + B is transferred to C. In a multi-transfer definition all assignments are made concurrently. VHDL differs from software programming languages in this respect; in a programming language, the order in which commands are written defines the order in which they happen, in VHDL the simulator assumes that all signal transfers are simultaneous, except where sequential processes are defined.

We can illustrate an RTL function definition with a multiplexer specified to operate as a discrete logic 157 type.

```
entity MUX157 is
    port( A, B: in BIT_VECTOR( 0 to 3);
        G, SEL: in BIT;
        Y: out BIT_VECTOR( 0 to 3));
end MUX157;
architecture RTL of MUX157 is
begin
    Y<= '0' when (G = '1') else
        B when (G = '0' and SEL = '1') else
        A
end ;
```

In order to specify state machines it is necessary to invoke sequential statements which occur within a process. The process itself is concurrent because it may be called at any time, when one of its signals changes. Essentially, a state machine consists of two parts, a sequential part which defines the ability of flip-flops to change state, and a combinatorial part which defines the signals offered to the flip-flops as a function of present state and inputs.

The door lock may be specified as follows:

```
entity DOOR_LOCK is
    port( OPEN, CANCEL in bit;
        CLOCK, RESET in bit;
        I in BIT_VECTOR (3 down to 0);
        UNLOCK, ALARM out bit);
end DOOR_LOCK
  architecture RTL of DOOR_LOCK is
    -- double hyphen is the VHDL comment syntax
    -- we must define internal signals which are not signal ports
    -- we do this by first defining types for the state signals
    type StateType is (S0, S1, S2, S3, S4, S5, S6, S7, S8, S9);
    -- then we define the signals we are going to use
    signal State, NextState: StateType;
    begin
        SEQUENCE: process( CLOCK, RESET)
        -- this defines a process called SEQUENCE which is
    invoked whenever
```

```
            -- there is a change in the value of CLOCK or RESET
        begin
            if (RESET = '0') then
                State <= S0;
                -- then we define a rising clock edge by
            elsif (CLOCK'event and CLOCK = '1') then
                State <= NextState;
            end if;
        end process;
            -- we can now define the state jumps as a combinatorial
    process
        COMBINATORIAL: process (I, OPEN, CANCEL, State)
        begin
            -- assign default output levels
            UNLOCK <= '0';
            ALARM <= '0';
            case State is =>
                when S0=>
                    if (I = 6) then
                        NextState <= S1;
                    elsif (I /= 6 and I /= 15) then
                        NextState <= S9;
                    end if
                when S1=>
                    if (I = 15) then
                        NextState <=S2;
                    end if
                    .
                    .
                    .
                when S9=>
                    ALARM <= '1';
                    if (CANCEL = '1') then
                        NextState <= S0;
                    end if;
            end case;
        end process    ;
end RTL;
```

This is the complete specification for the door lock function. A function of this size would normally form only part of a complete FPGA so the above would be just one entity of a total design listing.

Having defined a logic entity it may be used as often as desired within the whole design. Each use of the function is given a unique name and referred to this specification, a process called **instantiation** because each time the logic module is called up it forms an instance of that logic.

It is usual to verify each entity individually before connecting them together in the top level design. This requires a simulation routine to be devised and written, as described in the next section.

3.4.3 Simulation with VHDL

In order to simulate an entity, or a complete design, a test bench must be created. The syntax for this is similar to the syntax for specifying logic entities. For example, the entity declaration will be:

```
entity TEST_DOOR_LOCK is
end TEST_DOOR_LOCK;
```

This defines the test bench because it has no ports, only signal drivers. The architecture must define the signals, components and stimuli. It does this in the same way as in a logic architecture:

```
architecture TEST_BENCH of TEST_DOOR_LOCK is
    signal CLOCK: Std_Ulogic := '1'; -- defines '1' as the
      initial level
    signal RESET: Std_Ulogic := '0';
    signal OPEN: Std_Ulogic := '0';
    signal CANCEL: Std_Ulogic := '0';
    signal I: Std_Ulogic_Vector (3 downto 0) := '1111';
    signal DOOR: Std_Ulogic;
    signal ALARM: Std_Ulogic;
    constant CLK_PD: Time := 100 ns;
    constant RST_PD: Time := 50 ns;
    -- the Std_Ulogic type of signal allows bit definitions such
      as 'undefined'
    -- 'don't care', and 'tri-state' as well as '0' and '1', and
      give more information
    -- in a simulation result.
    component DOOR_LOCK
        port (
            CLOCK: in Std_Ulogic;
            RESET: in Std_Ulogic;
            OPEN: in Std_Ulogic;
            CANCEL: in Std_Ulogic;
            I: in Std_Ulogic_Vector (3 downto 0);
            DOOR: out Std_Ulogic;
```

```
        ALARM: out Std_Ulogic
           );
end component;
    begin
DOOR_LOCK
    port map (
        CLOCK → CLOCK, -- explicitly maps a signal to a port
           .
           .
           );
TB: BLOCK
begin
    CLOCK <= not (CLOCK) after CLK_PD/2; -- defines the 10 MHz
    clock
        RESET <= '1' after RST_PD;
        I <= 6 after 1 µs, 15 after 2 µs, 7 after 3 µs, 15 after 4 µs,
    1 after 5 µs -- etc.
           .
           .
end BLOCK TB;
```

Finally the results must be written to a file so that they can be examined to assess whether the simulation has achieved the desired result. VHDL includes a standard package, TEXTIO, to manage this function.

3.5 Schematic capture

3.5.1 Logic entry

Schematic capture is, probably, still the most popular method of defining logic for FPGAs, and many ASICs. It is a CAD system dedicated to logic design. Logic functions of complexity ranging from an invertor to multi-bit counters are stored in a library which describes both their functionality and a graphical symbol. The designer calls up the symbols from the library, places them on the screen of a PC or workstation, and connects them with wires and busses.

The design is taking place on two levels. At the visual level the designer is creating a visual representation of the logic which he requires, in terms of familiar symbols for the components. At a level below this is a netlist which defines the location of each component on the screen, and the way it is connected to the other components in the design.

Hierarchical design is still possible, indeed necessary, for all but the simplest of circuits. We can illustrate this with the door lock function which we have used through this chapter. Figure 3.3 shows how the state machine

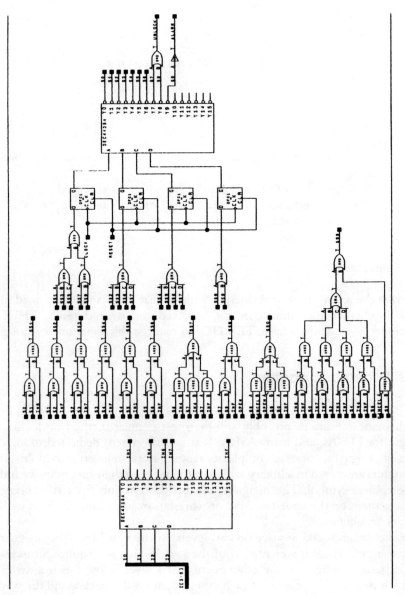

Figure 3.3 State machine schematic

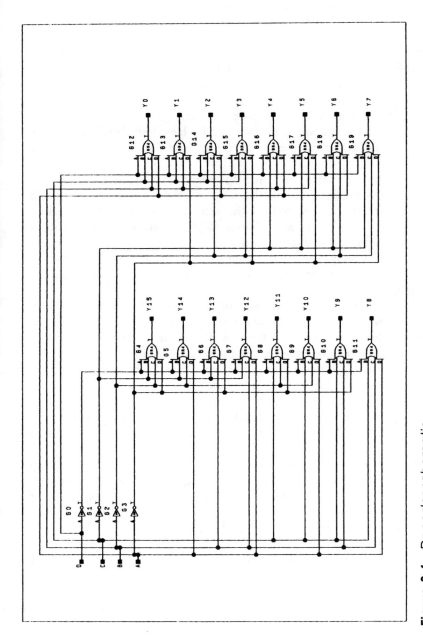

Figure 3.4 Decoder schematic

may be drawn up. This schematic was created using Viewlogic with the Actel cell library, but a similar result would be obtained with other capture packages and manufacturer's libraries.

The schematic may be broken into four parts which just fit onto a single sheet of the capture display. On the left is a decoder which is a standard Actel macro and generates the individual number inputs from the four-bit input line. The decoder schematic can be seen at a lower level of hierarchy; there is no need to draw the gate-level function on this sheet as the function of the block is clear. Replacing the symbol with the gate schematic, as in Figure 3.4, would make the picture less clear.

Note that the outputs from the decoder have been labelled IN1, IN4, etc. to indicate the number being input; note also that they are active-LOW.

A second decoder on the right-hand side generates signals to indicate the present state of the state machine. These are labelled S0, S1, S2, etc. to correspond to the state numbering we have already used. These signals are also active-LOW. Although the decoder function was pre-defined, by using it twice it only needs defining once. As a general point, any block of circuitry which is used in more than one location need only be defined once, but used in as many locations as desired.

The outputs from the state machine are generated from the state signals on the right-hand side of the drawing.

The block of gates in the middle of the schematic form the combinatorial section, defining the jump conditions from each state. They have been arranged to generate a 'next state' signal for each state. The jump to S0 is, in fact, superfluous since the state register is fabricated from D-types which set LOW in the absence of an input when clocked. It is included in our schematic for completeness although it would be automatically excluded at the place and route stage when components with 'dangling outputs' are eliminated.

Because the input and state signals are active-LOW, most of the gate inputs are 'bubbled'. Although a NOR gate with inverting inputs is logically equivalent to an AND gate, the bubbled input gates are used for clarity. For example, the top function says that state S1 is entered when a '6' is input in either state S0 or S1 (the hold condition).

The gate inputs are all labelled with the appropriate signal names. The connections do not have to be made with actual wires on the screen; if two nets are labelled with the same name they will automatically be connected in the netlist. The converse is also true; wires which must not be connected must have different names. Thus we have used 'S1' for a present state output but 'SS1' for a next state input.

The final section of the state machine is the state register and encoder. A full priority encoder is not needed to drive the state register because, if the

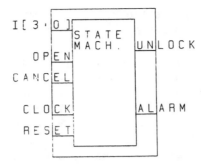

Figure 3.5 *State machine symbol*

logic is designed correctly, only one next state signal is active at any one time. Thus Q3 of the register must be set HIGH by SS8 or SS9, Q2 by SS4, SS5, SS6 or SS7 and so on.

The schematic was generated in the order in which it was described, except that a symbol was created first. This is shown in Figure 3.5, and could be used to define the state machine in a higher level of hierarchy; possibly, it could be incorporated into a single FPGA with a keyboard encoder, to form a complete system.

Part of the netlist is shown here. This is not intended to convey any information, except as an example of the structure used in saving the design information, and to show that while the drawing serves as a primary human interface, the netlist is in a form easily read by a computer.

```
DEF DOORLOCK; IN3, CANCEL, OPEN, IN0, CLOCK, RESET, IN2, IN1,
    UNLOCK, ALARM.
USE ADLIB: INBUF; $1I6.
USE ADLIB: INBUF; $1I5.
USE ADLIB: OUTBUF; $1I11.
USE ADLIB: INBUF; $1I9.
USE ADLIB: INBUF; $1I8.
USE ADLIB: INBUF; $1I7.
USE ADLIB: INBUF; $1I3.
USE ADLIB: CLKBUF; $1I10.
USE ADLIB: INBUF; $1I4.
USE ST_MACH; $1I2.
USE ADLIB: OUTBUF; $1I12.
NET $1N13; $1I11:D, $1I2:UNLOCK.
NET $1N15; $1I2:ALARM, $1I12:D.
NET $1N22; $1I2:OPEN, $1I7:Y.
NET $1N24; $1I2:CANCEL, $1I8:Y.
NET $1N26; $1I2:RESET, $1I9:Y.
```

```
NET $1N28; $1I2:CLOCK, $1I10:Y.
NET ALARM; ALARM, $1I12:PAD.
NET CANCEL; CANCEL, $1I8:PAD.
NET CLOCK; CLOCK, $1I10:PAD.
NET IO; $1I2:IO, $1I6:Y.
NET I1; $1I2:I1, $1I5:Y.
NET I2; $1I2:I2, $1I4:Y.
NET I3; $1I2:I3, $1I3:Y.
NET INO; INO, $1I6:PAD.
NET IN1; IN1, $1I5:PAD.
NET IN2; IN2, $1I4:PAD.
NET IN3; IN3, $1I3:PAD.
NET OPEN; OPEN, $1I7:PAD.
NET RESET; RESET, $1I9:PAD.
NET UNLOCK; UNLOCK, $1I11:PAD.
END.
```

3.5.2. Waveform generation and simulation

Having constructed a logic circuit, on paper, we must again show that it fulfils the desired function.

Just as we drew the circuit on the screen, at the same time generating an underlying netlist, so we can construct a set of test waveforms in 'oscilloscope format' and simultaneously produce the command file to drive the simulator. Figure 3.6 shows how we may exercise our doorlock.

We start by generating a 10 MHz clock and the power-on reset to initialize the circuit into state S0. The input starts at 'F'; changing it to '6' should change it to S1, and stay in S1 until the input changes back to 'F'. We continue the sequence which should run through the states until the door is unlocked, simulate the door opening and closing and, hopefully, return to S0.

Changing the input to '7' should set the alarm, in state S9, returning to state S0 on activating CANCEL. We can then go on and simulate any number of wrong entries — only one more is illustrated — to prove the complete functionality of the circuit.

The waveform may also be defined in a command file and checks included to ensure that the circuit operates as intended. The format for this is:

```
vector INPUT I[3:0]
| this defines the signals forming the input vector
vector STATES Q[3:0]
| this defines the state outputs
```

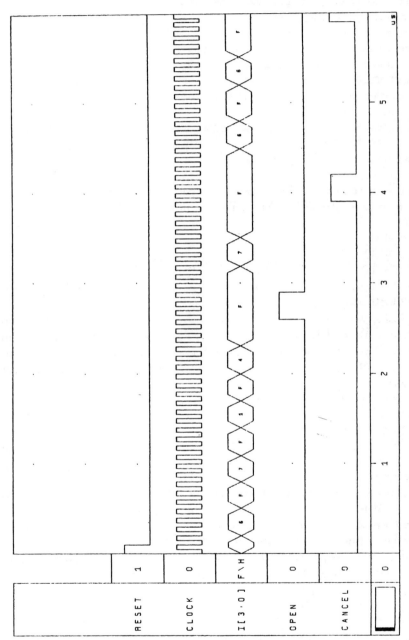

Figure 3.6 Waveform for PIN test sequence

```
vector SIGNALSIN RESET OPEN CANCEL INPUT
vector SIGNALSOUT UNLOCK ALARM STATES
| initialize the system
restart
wave DOORLOCK.WFM clock input states signalsin signalsout
| this defines the signals for display
wfm CLOCK 0=0 (500=1 500=0)*60
| this defines 60 cycles of 10 MHz clock (500 x 0.1 ns LOW, 500 x
    0.1 ns HIGH)
break CLOCK 0 do (assign SIGNALSIN < APPLIED; + check SIGNALSOUT
    < TESTPATT)
sim 60000
```

We then have to set up the applied vector pattern in file 'APPLIED' and the test pattern in file 'TESTPATT'. These will be as follows for APPLIED:

    4F\h
    0F\h
    06\h
    06\h
    0F\h, etc.

and for TESTPATT:

    >00\h
    00\h
    00\h
    01\h
    01\h
    02\h, etc.

The result of the simulation is shown in Figure 3.7.

The next step is to assemble the logic netlist into a real device, using the place-and-route program. Estimated timings may now be inserted into the netlist and the circuit resimulated to gauge its probable performance. In our case the circuit does not behave properly, however, if we slow the clock to 5 MHz as in the right-hand part of Figure 3.8, correct operation is restored. It appears that the circuit delays prove too great for the 100 ns clock period.

By zooming into part of the waveform we can obtain a good estimate of real circuit delays. Figure 3.9 shows the delay between the active clock edge and the alarm output; we can see that this is about 56 ns, for four logic levels plus input and output buffer. As the loop round the state machine is between six and nine levels deep, it is possible for problems to arise with a 100 ns clock period.

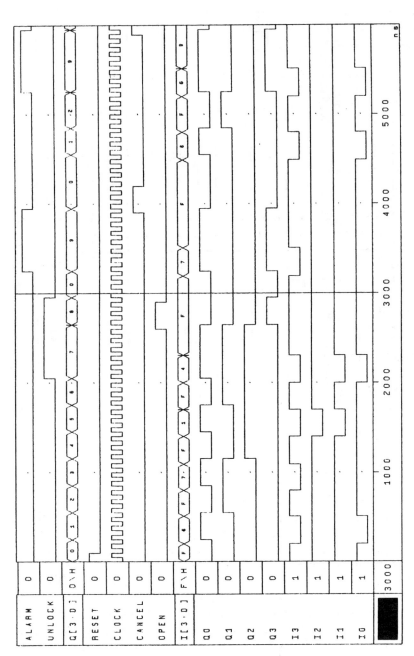

Figure 3.7 Display of simulation result for PIN detector

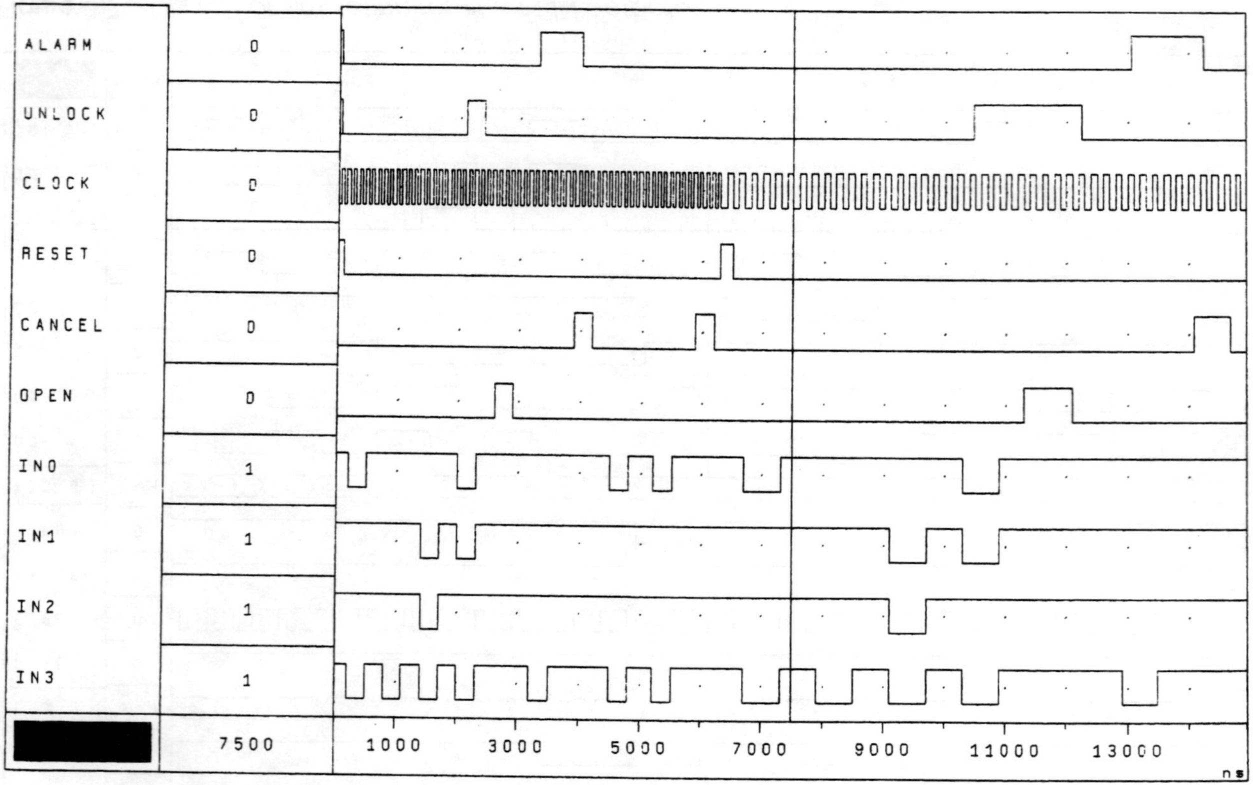

Figure 3.8 Post-layout simulation of PIN detector with 5 MHz clock

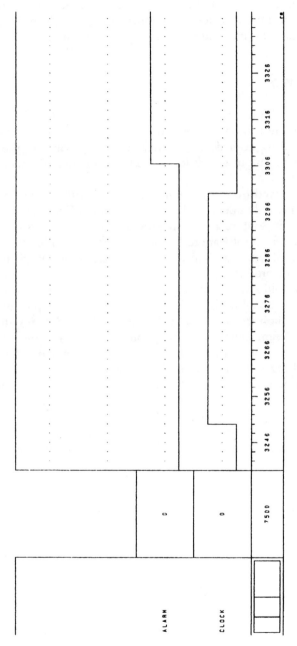

Figure 3.9 Expanded view of post-layout simulation

We shall see, in Chapter 8, how we can use a more efficient type of state machine which should reduce the number of logic levels in the loop and, thereby, increase the maximum clock frequency for state machines.

3.6 Conclusion

A variety of methods exist for defining the logic for FPGAs. As we noted, schematic capture is probably the most popular at present because this has been promoted by both device manufacturers and software providers. It also produces an output which looks most like a designer's mental picture of a logic system.

VHDL is derived from logic equation/state equation entry, which was the standard method of defining logic for 20/24-pin PLDs. However, it has a syntax which is closer to a software programming language so it should be easy to migrate from software to hardware design. It also has the benefit of being a universal language which is not targeted at any particular device manufacturer or product.

Both approaches have the capability of fitting a top-down design hierarchy. Both also allow for individual testing of the component modules before they are connected into a final structure. Most design systems cater for a mixed design approach, where some parts of the system may be defined in VHDL, the entities then being represented by symbols which can be connected together in a top-level schematic.

Whichever approach is used, the basic pattern is the same – logic entry, then logic simulation, followed by device definition and logic synthesis. Once the chip is laid out and routed, the estimated delays can be back-annotated to the simulation and real performance forecast. Proper use of the design tools will lead to a solution which meets the original design specification.

4 Large PAL structures

4.1 MACH families

4.1.1 First-generation architecture

The acronym MACH (macro array CMOS high-density) indicates the structure and process of this LSI programmable array logic (PAL) family from AMD. In terms of the speed/connectivity trade-off, MACH parts lie on the side of high speed. Examination of the block diagram of the MACH1 and MACH2 families, Figure 4.1, shows why this is.

Each device type has a small number of dedicated inputs, together with two to eight PAL-type blocks and associated I/Os. The '1' series have no buried macrocells, but the '2' series have as many buried macrocells as I/Os. Each family has three members, fitting into 44, 68 and 84-lead PLCCs respectively; thus the smallest MACH has 32 I/Os and macrocells, and the largest has 64 I/Os and 128 macrocells. In addition, each has six or eight dedicated inputs, some of which double as clock inputs.

The switch matrix allows only 26 inputs (22 in the two small MACHs) into each logic block. This is quite restrictive as there can be from 70 to 198 possible signals available at the switch matrix input although, in the largest MACH, only buried signals from the same or the adjacent block can be accessed. This restriction does mean, however, that every direct signal path from any input to any output is guaranteed to be 20 ns in the slowest MACH.

Each PAL block has four AND terms associated with each macrocell, but they are not necessarily dedicated to that macrocell. A block called the Logic Allocator assigns each group of four product terms to a macrocell, generally to the associated macrocell or one of its neighbours, although the '2' series has a slightly wider distribution. This function is shown in Figure 4.2.

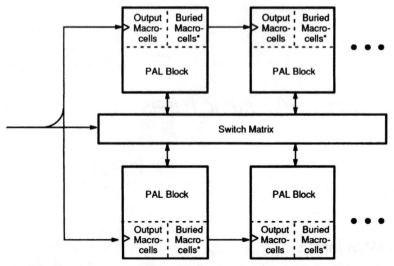

Figure 4.1 *MACH1 and MACH2 block diagram (reproduced by permission of Advanced Micro Devices)*

*Buried macrocell available on MACH2 devices only

Looking at the output macrocell, Figure 4.3, shows that it is very similar to the GAL macrocell described in Chapter 2. The flip-flop can act as a D-type or T-type or, in the '2' series, a simple latch. There are two multiplexers; one selects between combinatorial and registered outputs, the other between true and inverted signals. It is also possible to select which clock is used for each flip-flop in the

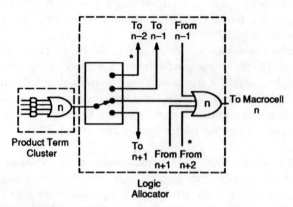

Figure 4.2 *MACH logic allocator (reproduced by permission of Advanced Micro Devices)*

*MACH2 only

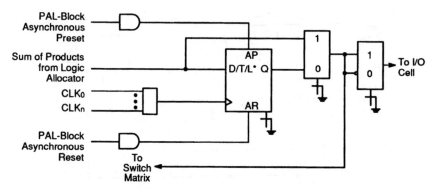

Figure 4.3 MACH output macrocell (reproduced by permission of Advanced Micro Devices)

Latch option available on MACH2 devices only

PAL block on an individual basis. Asynchronous preset and reset terms are also provided, but these operate across the whole PAL block.

An output is taken before the polarity multiplexer for feedback to the switch matrix while a second output, after the polarity multiplexer, feeds the I/O cell. This feature, shown in Figure 4.4, provides the interface between the logic blocks and the device pins, except for the dedicated inputs. The tri-state buffer can be set permanently open, permanently closed (when the pin is configured as an input) or controlled by one of two product terms which are common to all

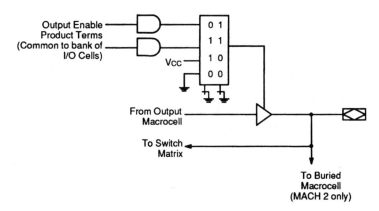

Figure 4.4 MACH I/O cell (reproduced by permission of Advanced Micro Devices)

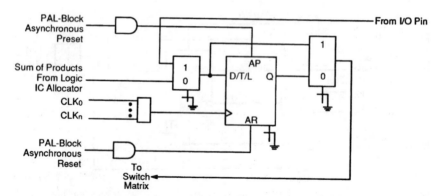

Figure 4.5 *MACH buried macrocell (reproduced by permission of Advanced Micro Devices)*

outputs in a PAL block. There are direct connections to the switch matrix and, in the case of MACH 2 devices, to the buried macrocell.

The buried macrocell is very similar to the output macrocell, as can be seen from Figure 4.5. There is no polarity multiplexer as polarity of feedback signals can be selected after the switch matrix, but there is a source multiplexer. This allows the input signal to be fed through the buried macrocell which then acts as an input latch or flip-flop.

The six MACH 1 and MACH 2 devices are designed for synchronous operation; all the internal flip-flops are clocked by one of the two or four dedicated clock inputs, but a seventh MACH has been introduced

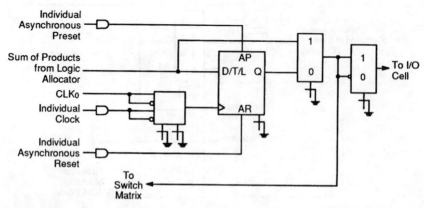

Figure 4.6 *MACH asynchronous macrocell (reproduced by permission of Advanced Micro Devices)*

specifically for asynchronous applications. Each macrocell has four extra product terms associated with it to provide individual clock, preset, reset and output enable from inputs from the switch matrix. The logic terms may be allocated to adjacent macrocells, in groups of four, as in the synchronous versions.

Figure 4.6 shows the asynchronous output macrocell. The main difference from the synchronous macrocell is the clock selection. Only one of the dedicated clocks may be selected, but either edge of the clock may be used; the individual clock is the other alternative, again with the positive or negative edge. Because the logic capability of any possible buried macrocells has been diverted to the generation of individual clocks and resets, the buried cells of the synchronous devices have been replaced by simple input macrocells which can use either edge of either clock to latch or register the input signals from the I/O cells.

This asynchronous device, the MACH215, has six dedicated inputs, two of which may be clocks, and 32 I/O pins in a PLCC44 package.

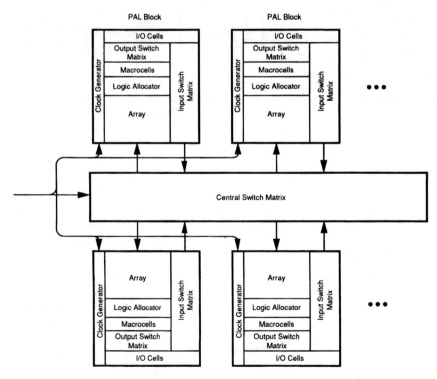

Figure 4.7 *MACH4 family block diagram (reproduced by permission of Advanced Micro Devices)*

4.1.2 Second-generation architecture

A second-generation of MACHs, the MACH 3 and MACH 4 families, has been introduced with higher I/O count, increased functionality and improved routing. In particular, there are additional features to increase the flexibility required when changes have to be made to existing designs.

The macrocells in these families are not associated with particular I/O cells; each PAL block has eight I/Os and eight or sixteen macrocells. An output switch matrix routes the macrocell outputs to the I/O cells. MACH 3 macrocells may be routed to one of eight I/Os, MACH 4 macrocells to one of four as there are twice as many macrocells as I/Os in this family. Flexibility in the opposite direction is provided by an input switch matrix defining which lines in the central switch matrix are driven by the various feedback signals from the macrocells and I/O cells.

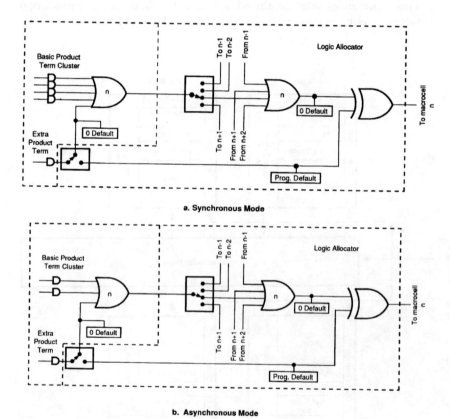

Fig 4.8 *MACH3 and MACH4 logic allocator (a) synchronous mode; (b) asynchronous mode (reproduced by permission of Advanced Micro Devices)*

Large PAL structures   73

The block diagram of part of a MACH 4 device is shown in Figure 4.7, which shows how the input and output switches fit in with the other blocks, familiar from the first-generation MACHs. One other 'new' feature is the clock generator. Each device has four clock inputs which are fed into a clock generator which distributes a programmable mix of true and inverted clock signals to the PAL blocks.

Additional logic capability is given by an extra product term per macrocell; it may be added to the basic four terms to give a group of five to the logic allocator, or it may be taken as a single term to one side of an exclusive-OR gate in the logic path. There could be two reasons for doing this; either to provide a simple logic function when the four-term cluster is allocated to a different macrocell, or to allow the macrocell to emulate J-K or T-type flip-flops.

The macrocells may be set, on an individual basis, as synchronous or asynchronous. In asynchronous mode, two of the product term clusters are

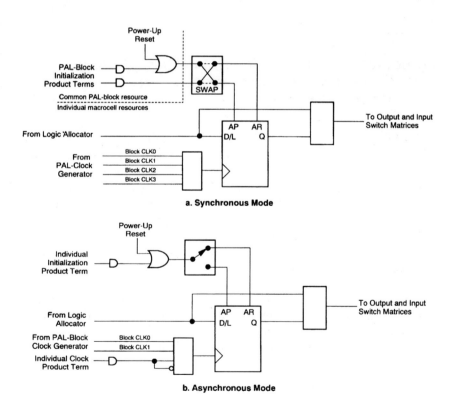

Fig 4.9 *MACH3 and MACH4 macrocell (a) synchronous mode; (b) asynchronous mode (reproduced by permission of Advanced Micro Devices)*

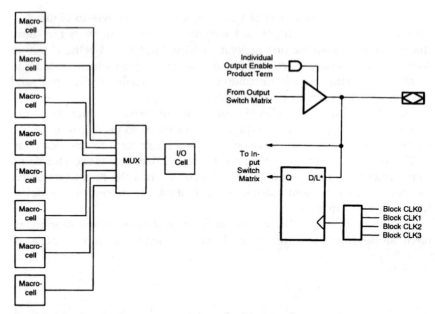

Figure 4.10 *MACH3 and MACH4 I/O cell (reproduced by permission of Advanced Micro Devices)*
*Flip-flop available on MACH4 devices only

used to provide the individual clock and initialization product terms. Initialization is combined with a power-up reset and may be connected to either the preset or reset inputs of the flip-flop. In synchronous mode, the preset and reset are provided on a global basis for each PAL block, the power-up reset being combined with either signal. The other difference between the first and second generation macrocells is that the input polarity is programmable in MACH 3 and MACH 4, rather than output polarity. The logic allocator and macrocell are shown, in both modes, in Figures 4.8 and 4.9 respectively.

The I/O cell in Figure 4.10, which also shows the output switch matrix, has two feedbacks to the input switch matrix; both the direct and a registered or latched signal are available. Each I/O cell has an individual output enable. The registered signal is available in MACH 4 devices only.

4.1.3 Device performance and design

As we noted above, the slowest MACH devices give a guaranteed maximum of 20 ns delay between any input and output, although selected devices as fast as 10 ns are also available. Maximum frequency of operation depends on many factors in a particular design but a 10 ns part can contain a synchronous counter capable of operation in excess of 80 MHz. This highly predictable delay is one of the most strongly noted features of the MACH family. It is due to all input and feedback signals passing through the switch matrix, which limits the loading on the logic terms by restricting the number of signals fed to each product term.

The other factor which influences many designs is the current consumption; this affects the design of the system power supply and overall cooling, and system reliability because of the chip junction temperature. Specified consumption ranges from 170 mA to 400 mA at zero frequency. Being CMOS, actual supply current depends on the frequency at which each circuit component switches and is, therefore, not very straightforward to calculate. The data sheets do include the incremental frequency-dependent values and allow an estimate of working consumption to be made.

Design data may be input by equations, schematic or synthesis tools, but there are some device-dependent features in the MACH design process. Fitting logic into simple PLDs is largely a question of not using too many inputs and outputs, and having enough product terms to fit the logic.

These criteria also apply to MACHs but there are the added complications of the switch array and logic allocator. The switch array limits the number of signals available to any PAL block so, if the 'wrong' outputs are grouped into a block and too many inputs are needed, the logic will not fit. Similarly, if two or more adjacent outputs need a high product term count, there may not be enough terms available for the logic allocator to provide for every output.

The best way to approach MACH designs is to allow the software to allocate I/O pins. It will try many possible combinations before selecting an optimum solution, assuming that one exists. If the fitter fails to complete a solution, it reports the problem areas and it may be possible to overcome these by rearranging the logic. One possible fix is to pre-combine some of the logic signals in a block that has spare input and logic resources. This will add an extra loop delay so it could not be used for critical paths although internal delay times are typically 2 ns less than input to output delays.

The technology used for MACHs is electrically erasable CMOS so, once programmed, devices can easily be re-programmed if changes need to be made. From the design point of view this is not usually much of a problem, provided that the changes do not involve bringing many extra signals into a PAL block, or adding significantly more product terms to outputs in areas where there is a high output density.

Those devices in packages larger than PLCC84 are designed to be programmed in-circuit with 5 V levels, avoiding mechanical damage to the MACH and electrical damage to the other circuit components. PLCCs may be programmed on most standard PLD programmers, although socket

Table 4.1 MACH device summary

Part number	Fixed inputs	I/O cells	Macro-cells	Av. Pts per cell	Max. Pts Ppr cell	Min.T_{pd} (ns)	Icc (0 MHz) (mA)
MACH110	6	32	32	4	12	12	150
MACH120	8	48	48	4	12	12	180
MACH130	6	64	64	4	12	15	180
MACH210	6	32	64	4	16	10	180
MACH220	8	48	96	4	16	10	300
MACH230	6	64	128	4	16	15	360
MACH215	6	32	32 (async)	4	12	12	180
MACH355	6	96	96	5	20	15	225
MACH435	6	64	128	5	20	12	255
MACH445	6	64	128	5	20	15	255
MACH446	6	64	128	5	20	10	255
MACH465	18	128	256	5	20	15	tba

adaptors may be needed for some of these. Programming is also covered in more detail later.

Table 4.1 summarizes the salient features of the MACH families.

4.2 MAX families

4.2.1 5000-series architecture

MAX (multiple array matrix) 5000 devices, introduced by Altera, conform to the general scheme of complex PLD architecture, in that they contain logic blocks with a central interconnection matrix linking them together, logically. Figure 4.11 shows the block diagram of a typical MAX 5000 device. The nomenclature of the various features is specific to the MAX family, but their function is no different from the equivalent part of any other complex PLD.

The interconnection matrix is called the PIA (programmable interconnection array); the logic block is an LAB (logic array block). There are some detailed architectural differences between MAX devices and MACHs. Dedicated inputs are fed directly to the LABs while I/O feedbacks, which could include direct inputs, are taken to the local LAB and the PIA. The macrocell feedbacks are also fed to the PIA and to their local LAB.

A novel function in the MAX is the expander array; this provides extra product terms for functions which require more than the three available in

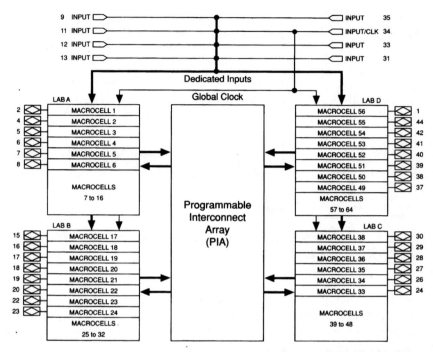

Figure 4.11 *EPM5064 block diagram (reproduced by permission of Altera Corporation)*

the standard macrocell. The expander array terms are product terms which are fed back to become further product term inputs. Each product term has sixteen inputs from the eight dedicated inputs (true/complement buffered), 48 inputs from the 24 allowed PIA signals, 48 inputs from local macrocells and I/Os and 32 expander inputs, a total of 144 compared with 44 to 68 in MACHs. There will be a speed penalty to pay for this but there is an increased flexibility in connectivity and I/O placement.

The flexibility of the macrocell may be judged from Figure 4.12.

Although there are only three pure logic terms an extra exclusive-OR gate, as found in second-generation MACHs, adds to the functionality. Each macrocell also includes its own product terms for clock, reset and preset inputs to the register element, and its own output enable. The register can be programmed as any of the standard flip-flop types, or as a latch, and can be clocked by the global clock or the locally generated clock.

Synchronous or asynchronous logic can be incorporated with equal ease into any of the macrocells, and may be mixed within LABs. Approximately half of the sixteen macrocells in each LAB are buried, and the dual feedback means that the macrocell associated with any I/O pin used as an input also

78 FPGAs and Programmable LSI: A Designer's Handbook

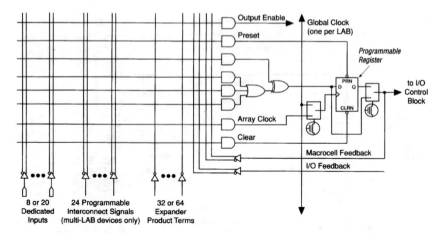

Figure 4.12 MAX 5000 macrocell (reproduced by permission of Altera Corporation)

becomes buried. The I/O pins themselves are standard tri-state buffers with individual control, as we have seen.

4.2.2 MAX 7000 architecture

The MAX 7000 family is an enhanced version of the MAX 5000 devices described in the previous section. The basic architecture is unchanged, with

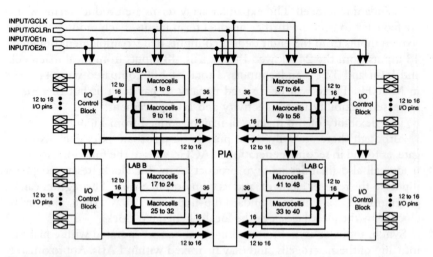

Figure 4.13 EPM7064 block diagram (reproduced by permission of Altera Corporation)

LABs connected via the PIA, but the local feedbacks have been eliminated except for the expander array. All signals, including direct inputs, I/Os and macrocell feedbacks, pass through the PIA. There are 36 inputs from the PIA to each LAB which now have 88 input lines to each product term, including the expander array. This is a significant reduction from the MAX 5000 situation so the performance is improved; this is at the expense of connectivity again. These features are summarized in Figure 4.13.

The MAX 7000 macrocell, shown in Figure 4.14, also shows some changes from the original family. Starting with the logic array itself, unused product terms from the macrocell below may be donated to the neighbouring macrocell. This is a better option than the expander array because there is a lower delay penalty in selecting an adjacent set of product terms than in making a pass through the expander array. The logic is also easier to define in this way because it is combined directly into the OR-gate instead of feeding a product term.

There are just five product terms per macrocell, compared with eight in the MAX 5000, but their allocation is not fixed. All five may be used as logic terms, connecting to either the OR gate or the exclusive-OR gate, or they may be used as control inputs to the output register element. This selection is made in the product term select matrix, which also accepts the parallel logic expander inputs. Global clock and clear signals are available if individual register control is not required.

The register can be set to any of the usual flip-flop types, or by-passed for combinatorial functions. Each flip-flop also has a clock enable input, which allows the designer to select individual flip-flops for clocking by the global

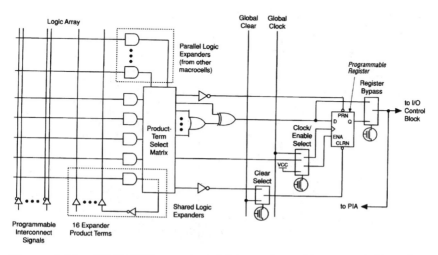

Figure 4.14 MAX 7000 macrocell (reproduced by permission of Altera Corporation)

clock input. It also simplifies the design and implementation of functions like counters which, otherwise, require separate definition of the 'hold' condition when using D-type flip-flops.

The I/O cell is very simple; it consists of a tri-state buffer, with four optional control inputs, and feedback to the PIA. The four control options are two global output enables, Vcc for direct outputs and ground for direct inputs. There is no provision for individual tri-state control.

4.2.3 MAX 9000 architecture

The basic macrocell of MAX 9000 is, to all intents and purposes, identical to the MAX 7000 cell shown in Figure 4.14; the difference between the two families lies in the way in which they are connected. In MAX 7000 all signals pass through a single PIA but in MAX 9000 each LAB has its own PIA, the individual PIAs being connected by a higher level of interconnect, called FastTrack. Figure 4.15 shows how the LABs fit into the FastTrack structure.

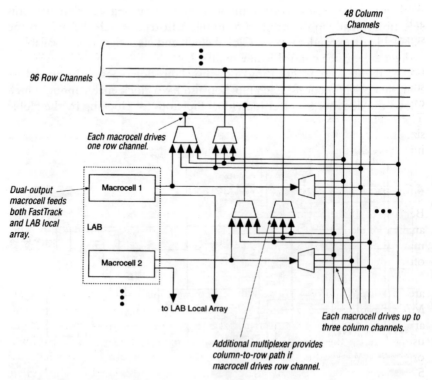

Figure 4.15 *MAX 9000 interconnect (reproduced by permission of Altera Corporation)*

Each row has 96 channels; each LAB PIA is fed by 33 signals from its associated row. With sixteen local feedback signals and sixteen shared expander terms the total width of the local PIA is 114. The sixteen outputs from the LAB are multiplexed with the column signals onto sixteen row channels; this path may, therefore, either route an output to the row tracks or turn a signal through a right angle. A second multiplexer may also be used for turning signals.

The vertical column contains 48 channels; each LAB output can drive up to three channels and, as we saw above, signals can be directed onto the rows at each crossing point. Signals can only be passed from a row to a column through a LAB.

Each row and column terminates in I/O cells, eight at each end of a row, ten at each end of a column. Outputs may be selected from one of ten row channels or one of seventeen column channels; this allows any LAB output to be routed to most of the I/O cells associated with its adjacent row and column. Each I/O cell can drive two channels when used as input.

The I/O cells themselves contain a flip-flop which may register either an input or output, and slew rate control for outputs. Control signals for tri-state, clock, clock enable and clear are derived from a peripheral control bus and may be sourced from the global control signals, or various array signals.

This hierarchical structure improves performance at a local level by restricting the PIA width; it is possible to build fairly complex functions, such as 16-bit counters, within a single LAB and use the high-speed routing capability of the FastTrack to connect these basic level functions together. The smallest MAX 9000 part, EPM9320, is equivalent to ten EPM7032s in size but, because the PIA is narrower and the LABs are connected internally, its performance is significantly better.

4.2.4 MAX performance and design

Because there are a number of different logic paths through MAX devices, analysis of delay times relies on a timing model for each device family. This may be performed by the design software or by a manual model, which is often useful for an initial appraisal of device performance.

As an example of the possible timing differences, two possible logic paths are shown in Figure 4.16(a), which is the timing model for a multi-LAB MAX 5000 device. The direct path from a dedicated input with one logic array delay and with a combinatorial output path has a delay time of 25 ns, using data for the EPM5064−1. An input using an I/O pin and involving one pass through the expander array gives a combinatorial logic delay of 52 ns. Most of this difference is accounted for by the delay through the PIA and expander array, but it shows how care must be taken when using the MAX 5000 family if skew is likely to be a problem.

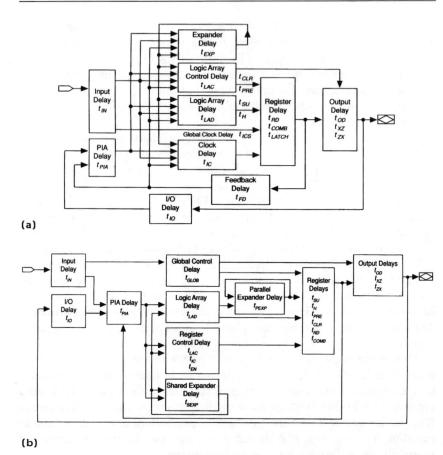

Figure 4.16 (a) MAX 5000 timing diagram (reproduced by permission of Altera Corporation); (b) MAX 7000 timing diagram (reproduced by permission of Altera Corporation)

A similar analysis based on the EPM7064−1 gives 10 ns for the most direct path, 10.8 ns if the logic path uses the parallel expander or 15 ns with the shared expander. This is clearly a much better situation than the first example, and the quoted operating frequency is 100 MHz compared with only 50 MHz for a standard MAX 5000 device configured as a counter. The timing model for MAX 7000 devices is shown in Figure 4.16(b).

Variability in the MAX 9000 family is more than MAX 7000 because of alternative paths through row and column FastTracks. Taking the fastest path from input to output in an EPM9560−15 gives a cumulative propagation delay of 16.0 ns. This assumes that the signal path is in and out via I/O cells connected to a row channel. As can be seen from the timing

Large PAL structures   83

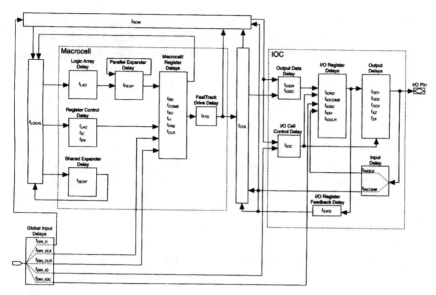

Figure 4.17 *MAX 9000 timing diagram (reproduced by permission of Altera Corporation)*

model in Figure 4.17, an input connected to a column has an extra delay because LABs must use a row as an input source. The output delay via a column is slightly higher also so, if input and output are both via column I/Os the propagation delay is 17.9 ns. This is not a vast difference but might cause problems in some applications where tight timing tolerances are required.

Supply currents range from 125 mA to 225 mA in the MAX 5000 family, from 35 mA to 225 mA for the MAX 7000 family and from 90 to 160 mA for MAX 9000 devices. As before these are zero frequency, zero load measurements and, as with all CMOS devices, power consumption rises with operating frequency. This dependency is characterized for each MAX device, although it will vary according to the function programmed into the part. Counters, for example, tend to be lower power than state machines whose flip-flops may run close to the clock frequency.

Each macrocell in the MAX 7000 and MAX 9000 series has a 'turbo' bit which, if set, allows the macrocell to run at full speed, otherwise it will run in a lower power mode. Only critical paths need the turbo bit to be set so the device may be optimized in speed/power terms.

MAX parts are quite flexible in terms of I/O assignment as there is not much interaction between adjacent macrocells unless, in the MAX 7000 family, a high product term count is needed for a particular output function. Again, connectivity is fairly high, particularly in the MAX 5000 family, so it unlikely that a LAB will run out of input resources from the PIA. It is, therefore, fairly reasonable to assign I/O functions to pins before fitting the logic into the device. The dedicated design system, MAX+PLUS II, will actually split a design into multiple MAX parts if it does not fit into a single device.

Programming presents no particular difficulties. The MAX 5000 family is built in EPROM technology so development devices may be obtained in window packages for erasing and reprogramming. MAX 7000 and MAX 9000, on the other hand, are EEPROM devices and may be reprogrammed in a standard programmer. In-circuit programming is only available for the MAX 9000 family, but high lead count packages in the other families are supplied in protective carriers to minimize lead damage before being installed.

Table 4.2 gives an overview of the basic features of the MAX families. Only Altera (EPM) part numbers are used, although some MAX parts are second-sourced under different part numbers. Some devices are available in more than one package – for example PLCC84 and QFP100. In these cases the I/O refers to the larger package style.

Table 4.2 *MAX device summary*

Part number	Fixed inputs	I/O cells	Macro-cells	Av. Pts per cell	Max. Pts per cell	Min. T_{pd} (ns)	Icc (0 MHz) (mA)
EPM5064	8	28	64	4	36	15	125
EPM5128A	8	52	128	4	36	15	225
EPM5130	20	64	128	4	36	15	250
EPM5192A	8	64	192	4	36	15	360
EPM7032	4	32	32	6	36	5	15
EPM7064	4	64	64	6	36	6	50
EPM7096	4	72	96	6	36	6	75
EPM7128E	4	96	128	6	36	7.5	60
EPM7160E	4	100	160	6	36	7.5	95
EPM7192E	4	120	192	6	36	10	110
EPM7256E	4	160	256	6	36	10	140
EPM9320	4	164	320	6	36	12	90
EPM9400	4	180	400	6	36	12	110
EPM9480	4	196	480	6	36	15	130
EPM9560	4	212	560	6	36	15	160

4.3 XC7000 EPLDs

4.3.1 XC7300 architecture

This family, from Xilinx, is unusual in that there are two different types of logic block with a different emphasis in each. One, aimed at high speed is called a fast function block (FFB), the other is the high density function block (HDFB). The general block diagram is shown in Figure 4.18. The universal interconnect matrix (UIM) accepts signals from the input pins, I/O pins and macrocell outputs for distribution to the function blocks, 24 to each FFB and 21 to each HDFB. If two or more inputs are directed to the same output line the signals are gated together, which provides an additional level of logic with no extra delay.

The FFB macrocell is shown in Figure 4.19. Twelve of the UIM signals are multiplexed with direct input lines and nine with feedbacks from the nine macrocells in the FFB. Each macrocell then has five product terms, four of which are ORed together and may also be combined with the four terms from the previous macrocell. This carrying forward is cumulative so that the ninth macrocell could have a function built from 36 product terms. The fifth term acts as a set to the output flip-flop, unless the block of four is being carried down, in which case it acts as a single logic term for the macrocell.

In most of the family the gated signals are inverted although the smallest CPLD (XC7336) has a programmable invertor at this point. It may also use the fifth term as a reset as well as a set.

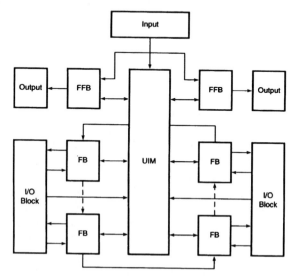

Figure 4.18 *XC7300 block diagram (reproduced by permission of Xilinx Corporation)*

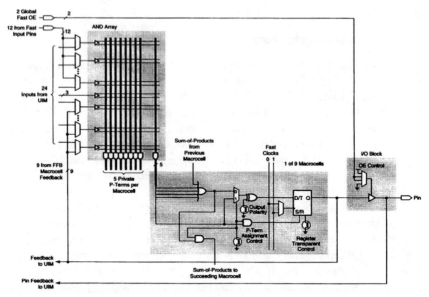

Figure 4.19 FFB macrocell (XC7300) (reproduced by permission of Xilinx Corporation)

The flip-flop may be D-type or transparent (combinatorial output), with a choice of one of the two Fastclock inputs. An additional option is available in the XC7336 which may also configure the flip-flop as a T-type. The output signal passes through a simple tri-state buffer with a choice of two global enables, or it may be set as a fixed output or input. This macrocell is much faster than the HDFB which has more switching possibilities within the macrocell as well as more complex functionality.

Figure 4.20 shows the high density function block (HDFB) macrocell. Each HDFB has three I/Os with fast inputs, and six I/Os which only feed the UIM; the three fast inputs, together with 21 lines from the UIM comprise the 24 inputs to the block. There are seventeen product terms available to each macrocell, five of these are dedicated to the cell while the other twelve may be used by any cell in the block. Four of the 'private' terms may be used as controls, set, reset, clock and output enable, or as logic terms where they may be combined with four of the shared terms in an eight-input OR gate. The fifth private term is combined with the remaining eight shared terms in a second OR gate; three of the shared terms may be replaced by local feedback or the output from the adjacent macrocell, which allows implementation of fast shift registers and similar functions.

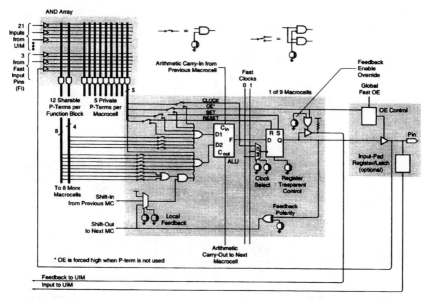

Figure 4.20 *HDFB macrocell (XC7300) (reproduced by permission of Xilinx Corporation)*

At the heart of the macrocell is an ALU, fed by the two OR gates. It may operate in either logic or arithmetic mode. In logic mode it generates any standard logical combination of the two inputs; as an arithmetic element it will produce the sum or difference of the two inputs and includes a carry in from the previous macrocell and a carry out to the next one. A single lookahead carry generator is provided for each HDFB to reduce the delay when all nine macrocells are used for arithmetic.

The output from the ALU is taken to a D-type flip-flop, which may be by-passed for combinatorial functions, and then to the I/O cell. Feedback from the flip-flop is available locally, via a programmable polarity buffer, and is directed to the UIM with optional control by the output enable.

Output control may be local or global, or a combination of the two, as shown in Figure 4.21. The output buffer may also be permanently disabled allowing the pin to be used as an input and the macrocell to become buried. The logic path from the pin to the UIM is through a selection of flip-flops and latches, or direct. One of the Fastclocks is used as the input clock or latch enable, and a separate clock enable may be used on one of the input flip-flops. The selection multiplexer is followed by a programmable polarity buffer.

With four potential levels of logic, this is the most complex architecture found in a large PAL structure.

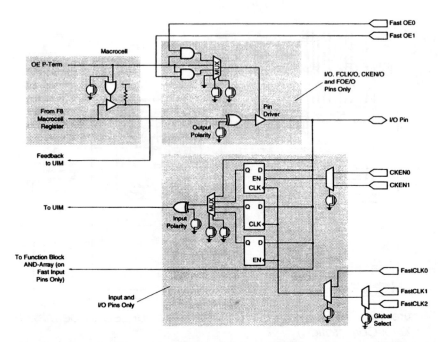

Figure 4.21 *XC7300 I/O cell (reproduced by permission of Xilinx Corporation)*

4.3.2 XC7200 architecture

The XC7300 was developed from the XC7200 which uses a similar macrocell to the HDFB. The major difference is the XC7272A in which the macrocell has less versatility than the XC7300 family. One of the private terms is dedicated to the clock; two further terms may be shared between set, reset and output enable, or gated with the two remaining terms and three shared terms. The second ALU input is generated from the nine remaining shared terms. Only 21 signals are available for generating product terms; these are all derived from the UIM as this device has no dedicated fast inputs.

The ALU functions are identical to the XC7300 described above, including the carry chain, but there is no shift chain or local macrocell feedback. There is also reduced functionality in the I/O cell with no global output enable or global clock enable inputs.

4.3.3 Application and performance

As well as a versatile architecture, some of the optional electrical features make this family flexible in its areas of application. While some of the other

CPLD families are becoming available in a 3.3 V only version, the XC7300 and XC7232A have an optional 3.3 V interface. This is achieved by separating the Vcc supplies for the internal logic and the output interface. While the internal logic must be run at 5 V, the output interface may connected to a 3.3 V supply. The input thresholds remain TTL and 3.3 V compatible but the output levels are pulled to the 3.3 V rail and may safely interface to a mixed I/O bus.

XC7300 Macrocells may also be run with low power or high speed options, thereby reducing unnecessary power consumption. Further reduction is achieved by turning off any macrocells which are not used. The typical supply currents range from 126 mA to 227 mA in low power mode, while the XC7200 consume from 126 mA to 252 mA.

A device as complex as this needs a timing model for estimating the delays through the different paths possible in the device; it is shown as Figure 4.22. As an example we can calculate timing for two combinatorial paths. The fastest path is from fast input through an FFB to a direct output; this is specified as tIN + tFLOGI + tFPDI + tFOUT. This comes to 10 ns for XC7354-10. A second typical path through the HDFB is tIN + tUIM + tLOGI + tPDI + tOUT, which evaluates to 21 ns, or 25 ns if the macrocell is set to the low power option.

Devices are fabricated using standard EPROM processing so only ceramic window packaged parts are erasable. In-circuit programming is not

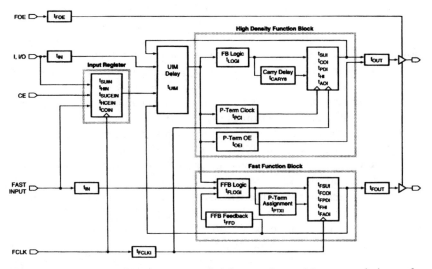

Figure 4.22 *XC7300 timing model (reproduced by permission of Xilinx Corporation)*

Table 4.3 *XC7000 device summary*

Part number	Fixed inputs	I/O cells	Macro-cells	Av. PTs per cell	Max. PTs per cell	Min. T_{pd} (ns)	Icc (0 MHz) (mA)
XC7236A	2	34	36	6.33	17	25	126
XC7272A	12	60	72	6.33	17	25	252
XC7318	17	18	18	5	36	5	90
XC7336	2	36	36	5	36	5	126
XC7354	9	49	54	5/6.33	36/17	7.5	140
XC7372	12	62	72	5/6.33	36/17	7.5	187
XC73108	12	108	108	5/6.33	36/17	7.5	227
XC73144	tba	tba	144	5/6.33	36/17	10	tba

available on current devices which must, therefore, be programmed before installation in a final assembly.

The salient features of XC7000 devices are listed in Table 4.3.

Where devices contain both FFBs and HDFBs the product term details of both function blocks are listed in the table. Icc values are typical for the low power option.

4.4 FLASHlogic

4.4.1 FLASHlogic architecture

FLASHlogic was introduced by Intel under the name FLEXlogic, but is now sourced solely by Altera who have another product called 'FLEX'. The new name is intended to represent the impending change of process to FLASH technology.

The basic structure consists of an interconnect matrix supplying input signals to configurable function blocks (CFBs). All inputs, I/O and CFB feedbacks are channelled through the interconnect matrix and, in addition, there are global clock inputs for driving the CFB flip-flops.

Each CFB has 24 inputs from the interconnect matrix, and these are allocated among the ten macrocells in each block. A separate 12-bit comparison circuit is fed by the same 24 inputs, grouped in pairs, providing a TRUE signal if the selected number of input pairs are all logically identical. This signal can be used as an input to one macrocell in each CFB, in place of one pair of product terms.

The product terms are generally arranged in pairs, two pairs per macrocell, but each pair may be diverted to the adjacent macrocell. The 'end' macrocells have an extra 10 terms permanently assigned to them

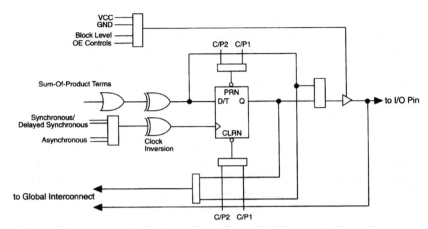

Figure 4.23 FLASHlogic macrocell (reproduced by permission of Altera Corporation)

giving a maximum of sixteen, compared with eight for 'centre' macrocells. The allocation scheme and macrocell circuit are shown in Figure 4.23.

In addition to the logic terms in each block there are six control terms providing two local clocks, two output enables and two set/reset signals. The global clocks are also available to each flip-flop, with or without a local delay, and the selected clock may be inverted, if required. The logic input also has a programmable inversion and may by-pass the flip-flop for combinatorial functions.

The macrocell output is via a tri-state buffer with the options of two control signals, Vcc (direct output) and ground (input only). The flip-flop output is also fed back to the interconnect matrix. An open drain option is available for the outputs; it may be used to implement wired-OR logic with more than one output connected to a single pull-up resistor.

A novel feature of the CFB is its ability to act as a block of RAM instead of logic. All ten macrocells have to be utilized and the configuration of one CFB is 128×10. Address, data and control signals are derived from the 24 input lines from the interconnect matrix. Control is as in a normal static RAM with a Block Enable, Write Enable and Output Enable signal, all active-LOW. Moreover, the initial data in the RAM can be preset by programming the block configuration cells, just as if it was being programmed with a logic function.

4.4.2 Design and application

Design of logic functions follows the same procedure as other PLDs. Other features may be defined by keywords or signal extensions in the design file.

Inputs may be set to interface to either TTL or CMOS levels; using the correct input definition will minimize standby current. Interfacing to 3.3 V CMOS requires use of the TTL interface, though, because the 2.4 V HIGH level is appropriate to the logic families operating at this supply voltage. Allocation of outputs to either 5 V or 3.3 V causes them to be connected to the appropriate supply pin; as mentioned above, there is also an open drain output option.

Logic timing is straightforward, thanks to the simple structure of the devices; that is, all inputs and feedbacks pass through the interconnect matrix. The three clock options do add a minor complication, however. Input setup times are 6 ns for the direct clock, 4.5 ns for the delayed clock and 2 ns for the asynchronous clock. Hold times are, respectively, 0 ns, 2 ns and 5 ns, as specified for the EPX740–10. The comparator circuit adds no delay time if it is just combined in the OR-array. To gate the comparator output with other signals it must be fed back, via a macrocell, to the interconnect matrix which incurs an extra propagation delay.

The RAM timing characteristics are specified separately, with the usual static RAM characteristics. For example, the EPX740–10 quotes a write pulse width as 10 ns with data setup and hold times of 10 ns and 2 ns respectively. For the read cycle, address access time is 15 ns and hold time 2 ns minimum from address change.

Typical standby currents are 1 mA for 'zero-power' options and 20 mA for the standard configuration.

The device programming is more versatile than most programmable logic. Each device holds configuration data in one of two areas, non-volatile cells and an SRAM block. If in-circuit programmability is required, the non-volatile cells are left unprogrammed and configuration data is loaded in the RAM at power-up, or at any time during normal operation. This allows easy prototyping changes to be made, or reconfiguration during

Table 4.4 FLASHlogic device summary

Part number	Fixed inputs	I/O cells	Macro-cells	Av. PTs per cell	x. PTs per cell	Min. T_{pd} (ns)	Icc (0 MHz) (mA)
EPX740	10	40	40	6	16	10	20
EPX740Z	10	40	40	6	16	10	1
EPX780	22	80	80	6	16	10	20
EPX780Z	22	80	80	6	16	10	1
EPX880	22	80	80	6	16	10	1
EPX8160	48	120	160	6	16	10	1

operation so that the same device can perform different functions according to the application.

The non-volatile cells may also be programmed in-circuit, or outside the circuit using a standard programmer. Once they are programmed, no changes can be made to the device configuration. Operation of the security bit is also determined by the non-volatile cell programming.

Table 4.4 compares the main features of the devices in the FLASHlogic family.

4.5 (is)pLSI families

4.5.1 (is)pLSI introduction

The Lattice LSI families are available with two programming options: ispLSI parts are in-system programmable, whilst pLSI devices must be pre-programmed on a stand alone programmer. There are three families, the LSI1000, 2000 and 3000. Although the families are architecturally similar

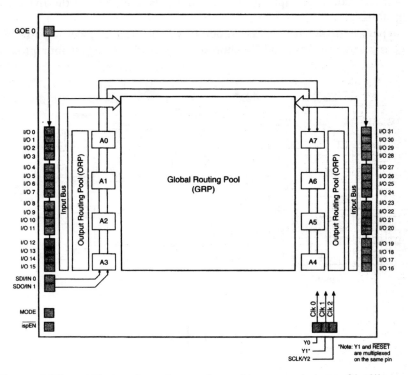

Figure 4.24 pLSI structure (reproduced by permission of Lattice Semiconductors)

there are differences in emphasis on speed and logic density, which are evident when their structures are examined in detail. These will become apparent in the next section.

4.5.2 (is)pLSI architectures

All three families share a common structure, as in Figure 4.24, with a GRP (global routing pool) at the heart distributing signals to and from the GLBs (generic logic blocks). The output signals from the GLBs are fed to the I/Os via the ORP (output routing pool). It is detailed differences in these three elements which lead to the performance differences in the three families.

The GLBs are virtually identical in each family. Looking first at the 1000 family, they consist of four macrocells with sixteen inputs available from the GRP, and two from dedicated input pins. Each macrocell is a typical PAL structure with two four-input OR gates, one five-input gate and one with seven inputs. However, a product term sharing array allows outputs from the OR gates to be allocated to any of the macrocells, in the manner of an FPLA. In addition, one of the product terms from each OR gate may be used, instead, as an input to an exclusive-OR gate driving the macrocell flip-flop.

There is an alternative, fast, mode in which the OR gate by-passes the sharing array to drive the flip-flop directly. In this case just four product

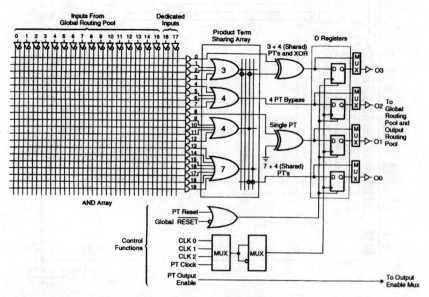

Figure 4.25 *GLB mixed mode operation (reproduced by permission of Lattice Semiconductors)*

terms are available, with two of the unused terms available as clock, reset or output enable signals. Figure 4.25 shows a GLB operating with a mixture of the various modes.

Eight GLBs are grouped together into a megablock which has sixteen I/Os, and two dedicated inputs associated with it. Outputs may be allocated in one of two ways; sixteen of the GLBs are directly connected to the I/Os, effectively by-passing the ORP for high speed operation, or all thirty two outputs are allocated via the ORP. Any GLB output may be allocated to one of four I/O pins; less than half the GLBs can be allocated to outputs, allowing for some I/Os used as inputs, so the remainder must use internal feedback.

The output enable is common for all I/O cells in a megablock and may be derived from an unused product term in any of the GLBs in the megablock. Alternatively, any I/O can be set as a permanent input or output.

All GLB outputs and I/O signals are fed to the GRP; inputs from I/O pins may be direct, latched or registered. The clocks for the GLBs and inputs are derived from a global clock network. Clock sources are either dedicated inputs or a dedicated GLB. These are multiplexed to provide three GLB clocks and two input clocks. The GLBs may also be clocked by a product term derived clock signal.

The clock distribution in the 2000 family is from dedicated inputs only, and no input registering is possible in the I/O cells. There is a global output enable available, as well as the GLB derived OE signal. The major difference between the 1000 and 2000 families is in the I/O capability.

Each megablock has two ORPs so every GLB is associated with an I/O pin in by-pass mode. Otherwise, they may be allocated to one of four I/O pins, as in the 1000 family. Although some of the I/O pins will be needed as dedicated inputs, many more GLBs can be given access to output pins than in the 1000 family.

The 3000 family has no dedicated inputs so all GLB inputs are from the GRP. In this family, the GLBs are paired off into twin GLBs and share 24 GRP inputs. 3000 family megablocks contain four twin GLBs, rather than eight single GLBs, but remain functionally identical. The I/Os themselves are also the same with the addition of boundary scan registers for JTAG compatibility. The clock network is also simplified with five dedicated clock inputs, three for GLB clocking and two for input registering.

4.5.3 (is)pLSI performance and programming

The device architectures allow designs to be optimized for speed or logic fit. All the details aimed at improving routability may be by-passed to minimize delay through the device. Signals by-passing the product term array and ORP suffer 10 ns delay compared with 13 ns when these features

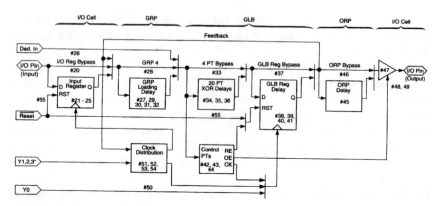

Figure 4.26 *LSI1016 timing model (reproduced by permission of Lattice Semiconductors)*

are included. Device specifications include timing models, as in Figure 4.26 for the LSI1016, which allow detailed prediction of the delay through any path in the device to be made.

Supply currents are, typically, in the 100–200 mA range as with most CPLDs; there is no zero-power option. The main device features are summarized in Table 4.5.

Devices may be programmed in-system or on a dedicated programmer, off-board. ispLSI parts have a five wire interface for programming after installation onto a circuit board. The programming source may be a remote PC or workstation, or a local intelligent controller. Target devices may be programmed in parallel, with separate enables, or daisy-chained. It is

Table 4.5 *(is)pLSI device details*

Part number	Fixed inputs	I/O cells	Macro-cells	Av. PTs per cell	Max. PTs per cell	Min. T_{pd} (ns)	Icc (0 MHz) (mA)
LSI1016	4	32	64	5	20	10	150
LSI1024	6	48	96	5	20	12	190
LSI1032	8	64	128	5	20	12	190
LSI1048(C)	10 (12)	96	192	5	20	15	235
LSI2032	2	32	32	5	20	7.5	40 typ.
LSI2064	4	64	64	5	20	7.5	tba
LSI2096	6	96	96	5	20	10	tba
LSI3192	0	96	192	5	20	10	tba
LSI3256	0	128	256	5	20	15	150 typ. tba
LSI3320	0	160	320	5	20	15	tba

possible to use all programming interface pins as normal device inputs by multiplexing system signal lines with the programming inputs. An instruction to the contrary during the routing phase will keep the programming interface separate.

Because the LSI parts use EEPROM technology they need be programmed only once, whether on- or off-board. However, their reprogrammability means that they may be upgraded without being removed from the board, making system changes possible with minimum upheaval.

Lattice offer the pDS design system for Boolean entry, device fitting and simulation. LSI parts are also supported by many third-party design systems, usually with an interface to pDS for the place-and-route operation. Standard programming files are generated for standalone programming or isp programming.

4.6 LSI sequencers

4.6.1 PA7140

The PA7140 is the largest part in a family, manufactured by ICT, which includes 24- and 28-pin parts. It has 38 I/Os and 24 logic cells. The logic

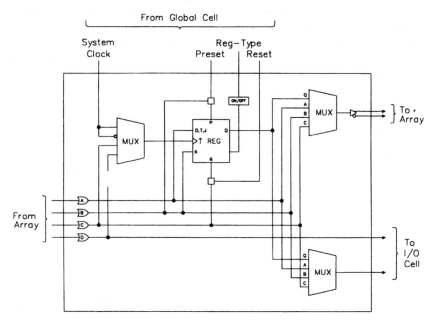

Figure 4.27 *PA7140 logic cell (reproduced by permission of ICT Corporation)*

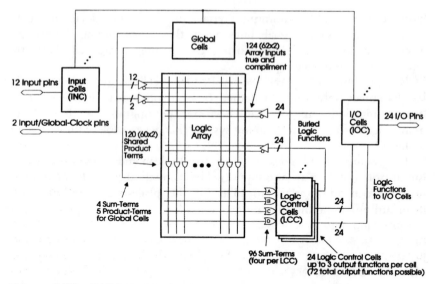

Figure 4.28 PA7140 structure (reproduced by permission of ICT Corporation)

cell, shown in Figure 4.27, has four OR gate inputs which may act as flip-flop inputs, flip-flop clock, set or reset, combinatorial feedback or combinatorial output. Array feedback may also be taken from the flip-flop output, which may be taken to an output as well.

Figure 4.28 shows that the PA7140 has a typical PLS structure. The product terms are split into two banks of sixty to reduce loading on the OR array and help maintain speed. All inputs and feedback terms are available to every product term so it is relatively straightforward to use the PA7140 to build either Moore or Mealy state machines. Because the flip-flops can be configured to J-K mode the jump conditions can be mapped directly onto the AND array.

The PA7140 can be clocked at up to 60 MHz with a maximum supply current of 150 mA. Design is supported on ICTs PLACE development system, and by most independent design suites. The devices are fabricated in EECMOS, so they are non-volatile but reprogrammable, making them a useful tool for development as well as production.

4.6.2 PML2552/PML2852

PML devices, from Philips, are based on a folded NAND array. The usual sum of products structure may be written as:

Y = A & B # C & D

Large PAL structures 99

which is logically equivalent to:

Y = !(!(A & B) & !(C & D))

Figure 4.29 shows how these two forms may be represented schematically. By folding the NAND outputs back to the inputs of a third NAND gate, two-level logic can be performed on a single level array.

The PML2552 and 2852 have 96 NAND gates with outputs folded back. These parts also have 29 inputs, 16 of them registered, 24 bidirectional I/Os and 20 flip-flop true/complement feedbacks making 242 fuse inputs to each NAND gate. This high number of programmable cells has a strong effect on the speed of a gate; the worst-case delay through an internal NAND gate is 20 ns.

The overall structure, depicted in Figure 4.30, shows how cells which might normally be driven by an OR array, fixed or programmable, are just added alongside the NAND gates which are performing the combinatorial logic. Both devices have two blocks of J-K flip-flops with internal feedback; one block is ten wide with a common clock, the second is ten wide with individual clocking. The clocks are derived from four external pins, one common with the input register, four dedicated NAND gates or the outputs of the second block of flip-flops.

Sixteen of the bidirectional I/Os are registered with by-passable D-type flip-flops, while all 24 I/Os have either individual or a common tri-state enable. The PML2852 has an extra 16 dedicated output pins with enables common in banks of four, or common with the other I/Os.

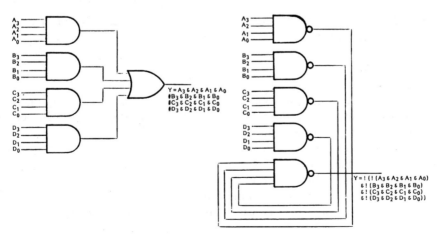

Figure 4.29 *AND/OR - NAND/NAND equivalence*

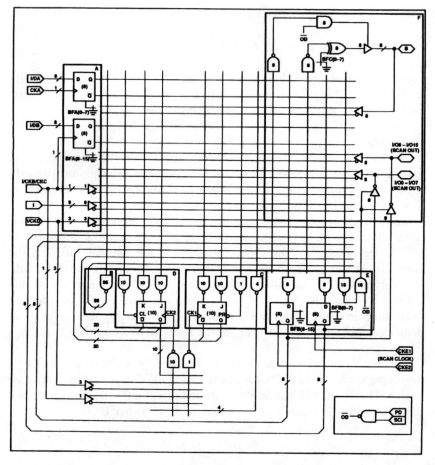

Figure 4.30 *PML structure (reproduced by permission of Philips Semiconductors)*

Scan path testing is available on both devices, but not conforming to JTAG specification. It does allow all internal flip-flops to be loaded, exercised and then observed.

The devices are fabricated in a CMOS EPROM process so reprogrammability is only possible with windowed parts. Design is based on Philips SNAP program, which includes a device compiler and delay-dependent simulation. There is also a limited interface from some schematic capture and third party Boolean entry packages.

Table 4.6 lists a summary of sequencer details.

Large PAL structures 101

Table 4.6 *LSI sequencers*

Part number	Fixed inputs	I/O cells	Fixed outputs	Product terms	Sum terms	Flip-flops	Min. T_{pd} (ns)	Icc (0 MHz) (mA)
PA7140	14	24	0	120	96	24	20	150
PML2552	29	24	0	96	95	36	55	10
PML2852	29	24	16	96	103	36	55	10

4.7 FLASH370 series

4.7.1 FLASH370 architecture

Recently announced by Cypress, and occupying a position midway between CPLDs and LSI sequencers, the FLASH370 series offers a high speed and relatively dense family of LSI programmable logic.

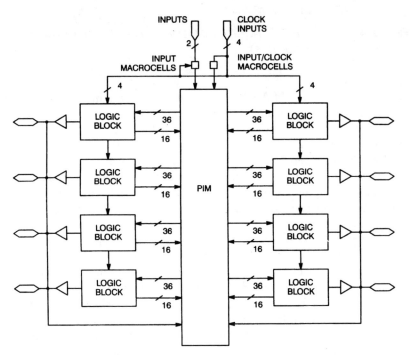

Figure 4.31 *FLASH370 block diagram (reproduced by permission of Cypress Semiconductors)*

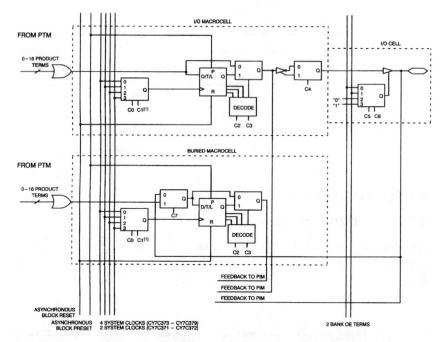

Figure 4.32 *FLASH370 macrocell (reproduced by permission of Cypress Semiconductors)*

Like many CPLDs, FLASH370 is based on a central interconnect matrix, the PIM (programmable interconnect matrix), taking signals from I/Os and macrocell feedbacks, and distributing them back to the macrocells. The macrocells are arranged in logic blocks, each with sixteen macrocells and from eight to sixteen I/Os, depending on the device.

Each block is fed with 36 signals from the PIM and generates eighty logic product terms, plus six more for control functions. The logic terms are distributed to the individual cells by a product term allocator; up to sixteen terms may drive one macrocell. Product terms may also be shared by up to four macrocells in quasi-PLA fashion, making it appear like a small sequencer.

Each macrocell contains a by-passable flip-flop which can be D-type, T-type or latch. Both buried and visible macrocells pass the registered/by-passed signal back to the PIM; visible macrocells drive a tri-state output buffer, controlled by one of two product term generated enables. The I/O pin also drives into the PIM. Each flip-flop has a choice of four clock lines, derived from the four input/clock pins. The other control terms provide asynchronous set and clear for all the flip-flops in the block.

Each device has six dedicated inputs, up to four of which can provide clocks, and may be latched or single or double registered as inputs. I/Os used as inputs can be registered in their associated buried macrocell, if any, but this prevents the macrocell from being used as a logic cell.

The block diagram and macrocell structures are shown in Figures 4.31 and 4.32 respectively.

4.7.2 FLASH370 performance

Timing analysis for this family is quite straightforward. Specified timings take all possible configurations into account, including product term sharing and maximum product term allocation. Thus the input to output delay for the smallest part, CY7C371, is 8.5 ns maximum for any single pass through the device.

Power consumption is typically from 100 mA to 300 mA, depending on the device complexity, with no zero-power option. The family features are summarized in Table 4.7.

Table 4.7   FLASH370 family

Part number	Fixed inputs	I/O cells	Macro-cells	Av. PTs per cell	Max. PTs per cell	Min. T_{pd} (ns)	Icc (0 MHz) (mA)
CY7C371	6	32	32	5	16	8.5	175
CY7C372	6	32	64	5	16	10	250
CY7C373	6	64	64	5	16	10	250
CY7C374	6	64	128	5	16	12	300
CY7C375	6	128	128	5	16	12	300
CY7C376	6	128	192	5	16	15	tba
CY7C377	6	192	192	5	16	15	tba
CY7C378	6	128	256	5	16	15	tba
CY7C379	6	192	256	5	16	15	tba

FLASH370 devices will fit into the same socket as some of the MACH family described in Section 4.1, although the architectures are not identical. By using VHDL, it is possible to target either family with the same design data. Cypress's Warp2 and Warp3 design systems both support VHDL, while Warp3 interfaces to third-party schematic capture and simulation tools.

As its name implies, FLASH370 is fabricated in FLASH technology, so the devices are non-volatile, but re-programmable electrically.

5 RAM-based FPGAs

5.1 LCA families

5.1.1 Overview

LCAs (logic cell arrays) were the first FPGAs to be introduced, in 1985, with the XC2000 family from Xilinx. Since then three new families have been added to the stable, the XC3000, XC4000 and XC5000, together with some speed and power supply variants. These four generations of FPGA use the same basic architecture, as described in Chapter 2, but each is enhanced from the previous issue. We will consider them as a group so that the enhancements can be highlighted against the background of the common architectural features.

All four families offer a relatively coarse-grained structure with from 64 to 1024 logic cells across the range of 600-gate to 25 000-gate devices. The cells have from four to nine inputs and two or four outputs. Each cell is set in a matrix of horizontal and vertical routing channels which connect to the I/O cells at the periphery. Some additional peripheral logic is available in the XC4000 family.

The smallest devices are offered in a 44-lead PLCC, with 34 I/Os, while the largest come in 299-lead pin-grid arrays with 256 I/Os. The next sections examine the features in more detail.

5.1.2 CLBs

The LCA cells are called CLBs (configurable logic blocks) and, in all three families, contain a combinatorial block driving one or two flip-flops. Figures 5.1 to 5.4 depict the CLBs for the XC2000, XC3000, XC4000 and XC5000 families, respectively.

Taking the XC2000 first, the combinatorial block has four inputs, one of which can be feedback from the flip-flop, and two outputs. Each output

RAM-based FPGAs 105

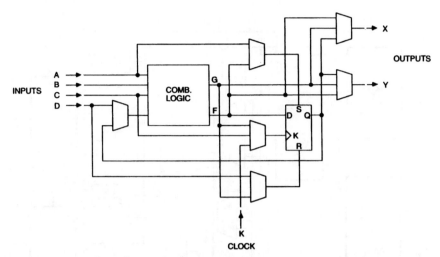

Figure 5.1 XC2000 CLB (reproduced by permission of Xilinx Corporation)

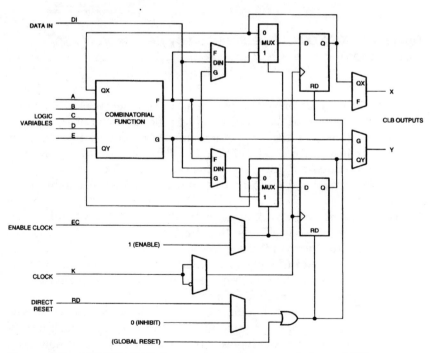

Figure 5.2 XC3000 CLB (reproduced by permission of Xilinx Corporation)

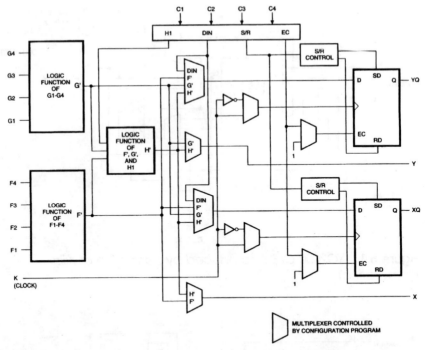

Figure 5.3 XC4000 CLB (reproduced by permission of Xilinx Corporation)

can be the same function of the four inputs or an independent function of three of the inputs. One output drives the flip-flop D-input or set input, or one of the cell outputs direct, the other output is an optional flip-flop clock or reset, or may drive a cell output. The flip-flop may be set, reset or clocked by a designated cell input or clocked from a global clock line. The cell outputs may be the flip-flop output or one of the combinatorial functions.

The various logic paths through the CLB are defined by multiplexers which are, in turn, controlled by RAM cells embedded in the CLB. The combinatorial function is a PROM-like structure capable of generating any Boolean function; again, embedded RAM cells define the function. By multiplexing between the two three-input structures, it is also possible to implement some five-input logic functions.

This structure is extended in the XC3000 family to a five input function block with two flip-flops. The two outputs of the function block may be independent functions of four inputs or the same function of five inputs, two of which may be feedback from the flip-flops. The independent outputs may be multiplexed to give some six or seven input logic functions.

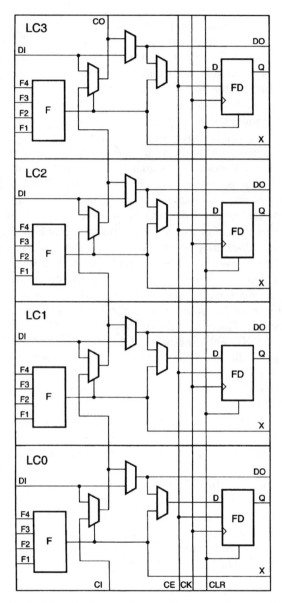

Figure 5.4 *XC5000 CLB (reproduced by permission of Xilinx Corporation)*

They may be used as direct outputs from the CLB, or clocked by one of the flip-flops.

Instead of the combinatorial functions the flip-flops may use a direct data input, which by-passes the logic block, or, when the enable is LOW, held in its present state by feeding the output back to the input. Both flip-flops share a reset and invertible clock and are multiplexed into the CLB output with one of the combinatorial functions.

Even more complexity is achieved in the XC4000 CLB. Two logic blocks, each with four independent inputs feed a third block where they may be combined with a ninth input. Either a simple function or a combined function is available from two combinatorial outputs, while any of the block functions or an independent data signal may feed either flip-flop. These have their own output making four outputs from each CLB. Each flip-flop has a common control line, individually configurable as set or reset, a common enable with individual override and an invertible common clock.

An important feature of the XC4000 function blocks is their arithmetic capability. Each block may be configured as a 2-bit adder with fast carry. The carry circuit may be extended over sixteen bits without degrading speed, making this family ideal for fast arithmetic operations.

The carry network may also be used to build fast counters; a 16-bit up/down counter may be built from eight CLBs and run at 40 MHz in XC4000 devices. The same function occupies 27 CLBs for 30 MHz operation in the XC3000 family.

A further possibility with XC4000 is to configure the two larger function blocks as 32-bit RAM.

Access time is the same as the CLB propagation delay, about 5.5 ns. This embedded RAM can form the basis of very fast on-chip FIFOs and registers, leading to many novel designs which are not possible with external memory.

The XC5000 family takes a step backwards in complexity, inasmuch as the basic cell is a four-input function generator and flip-flop with simpler interconnection than the XC2000 CLB, even. However, the technology used, three layer metallization, means that the CLB, or VersaBlock, can consist of four simple cells with very short, local routing resources. Each CLB thus has twenty inputs, twelve outputs and a fast carry network. Thus, although each basic cell is very simple the graininess, as seen by the global routing, is coarser than any of the other families.

5.1.3 I/O blocks

I/O blocks are the interface between an FPGA and the rest of the system and, in common with most FPGAs, offer the designer many features to simplify this process. We will not describe each family in detail because most of the features in each family are common.

Most of them offer CMOS/TTL input threshold selection, tri-state and slew rate controls for the outputs, and optional pull-up and pull-down resistors. The XC3000 and XC4000 families have flip-flops/latches for the input and output lines, XC2000 has an input flip-flop while the XC5000 relies on adjacent CLBs for signal registering, if required.

An additional feature of the XC4000 families which is located on the chip periphery, although not strictly an I/O function, is the Wide Edge Decoder. There are four on each FPGA and they vary the width through the family from 42 to 72 inputs. They may also be split to provide a larger number of narrower decoders. They are intended to compete with the advantage that PALs have with their wide logic gates. Decoding a complete address bus may need two or three levels of gating in an FPGA, making them slower than PAL-type devices which can decode in a single level. The Wide Edge Decoder enables the XC4000 family to match PAL speeds in this area.

One problem sometimes faced in FPGAs is that I/Os need to be physically close to the CLBs which are using or providing those signals. Many place-and-route programs will only work efficiently if the I/Os are not pre-placed, allowing the software to minimize signal tracking around the chip periphery. This is not necessarily a problem with the initial design, but may become troublesome if changes have to be made. At this stage, the device pinning may have been fixed to allow PCB design to go ahead, but the layout program may need to change the pinning to accommodate logic changes.

The XC5000 family has a VersaRing surrounding the logic blocks. This has additional routing and switching resources and makes it easier to route between I/O blocks and the logic matrix.

5.1.4 Cell interconnection

As with the CLBs, cell interconnection for the three families shows an evolution from XC2000 through XC3000 to XC4000 and XC5000. The simplest, and earliest scheme, for the XC2000, is shown in Figure 5.5. Horizontal and vertical channels between the CLBs contain, respectively, four and five short lines, and one and two long lines.

The short lines enter a switch matrix at each crossing point, where they can be connected to one or more of the other lines terminating there. In this way, logic signals are distributed in both directions around the FPGA. There are also buffers at intervals in the routing channel to overcome the problems of track and switch capacitance for signals travelling any distance.

The long lines by-pass the switch matrices and are intended for carrying signals over long distances with high fan-out and low skew. A global long line carries the clock for direct connection to the clock input of any CLB.

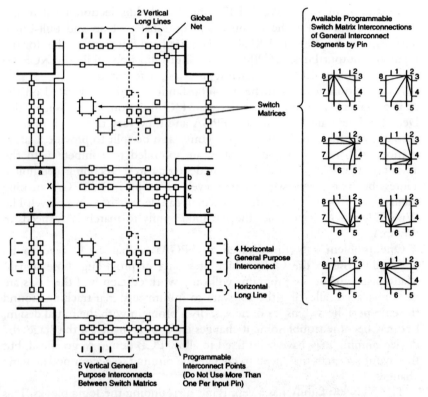

Figure 5.5 *XC2000 interconnect scheme (reproduced by permission of Xilinx Corporation)*

The fastest signal path is provided by short fixed connections which join each output to some of the inputs of adjacent CLBs. Direct connections are also provided between the I/O cells and peripheral CLBs.

Figure 5.6 shows a detailed view of one corner of an XC3000 chip. Connections to CLB inputs and outputs are via PIPs (programmable interconnection points), one side being connected to a CLB I/O, the other to one of the interconnect resources.

There are five short lines in each XC3000 routing channel, with a switch matrix at each crossing point, as before. As in the XC2000, there are regular bi-directional buffers to prevent signal degradation on long nets. Direct connections between adjacent blocks are also provided.

The long lines, three vertical and two horizontal, plus the global net, perform the same function as before. The horizontal long lines have an additional feature; they can be driven via tri-state buffers with an on-chip pull-up resistor. It is possible, therefore, to imple-

RAM-based FPGAs 111

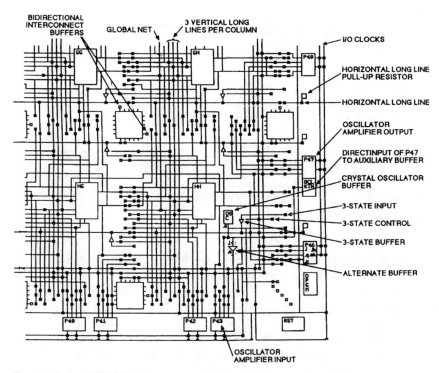

Figure 5.6 *XC3000 interconnect detail (reproduced by permission of Xilinx Corporation)*

ment standard bus functions like wire-AND directly within the FPGA.

Additional features in the XC4000 interconnect include symmetrical placement of CLB I/Os, double length lines and a simplified switch matrix. The performance improvement in the switch matrix makes direct connections unnecessary.

The other LCA families had the outputs on the right-hand edge of the CLB disposing data flow in general to be from the left-hand edge of the chip to the right. XC4000 have inputs and outputs on all four edges which makes placement and routing more like conventional ASICs. Local signals are usually routed along single length lines; the double length lines have a switch matrix at alternate crossing points and are used for intermediate length signals.

As with the XC3000 there are long lines in both directions, with the horizontal lines capable of performing bus functions. The long lines have switches at the mid points so that they can form independent half length connections.

112 FPGAs and Programmable LSI: A Designer's Handbook

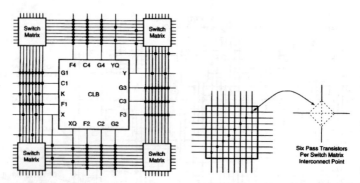

Figure 5.7 XC4000 single length lines (reproduced by permission of Xilinx Corporation)

The general XC4000 interconnect capability is shown in Figures 5.7 to 5.9. CLBs and I/Os may be connected to any of the levels, but only single length and long lines may actually be joined to each other where they cross. The 'global' long lines, with high-drive buffers, drive the clock and CLB control inputs.

The XC5000 family also has single length, double length and long lines in its General Routing Matrix, but uses the third metal layer for local

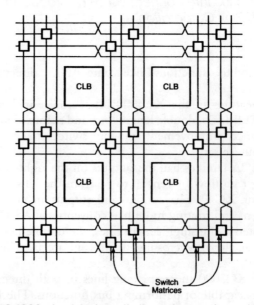

Figure 5.8 XC4000 double length lines (reproduced by permission of Xilinx Corporation)

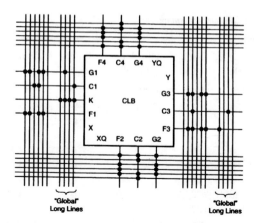

Figure 5.9 *XC4000 long line routing (reproduced by permission of Xilinx Corporation)*

routing. This includes routing between cells in a VersaBlock and routing between adjacent VersaBlocks. Fairly complex logic functions can be built from neighbouring CLBs and connected with the fast local tracks. These distinct logic blocks can then be interconnected via the global routing resource. This hierarchical approach to layout mirrors the top-down design strategy described in Chapter 3.

However fast the cell logic might be, in an FPGA, performance may hinge on the capability of the interconnection resources. With three levels of interconnect in all the families, these LCAs should have enough resource to ensure that connection issues do not become a problem. In particular, the symmetrical layout of the XC4000 CLBs makes routing more versatile and minimizing track lengths much simpler, while the local connections and VersaRing in the XC5000 increase the flexibility and performance of designs made in this family. This makes a high performance solution highly probable.

5.1.5 LCA range and performance

The LCA range of XC2000, XC3000, XC4000 and XC5000 is, in fact, eleven FPGA families although there are just the four basic architectures described above.

The XC2000 architecture has two versions, XC2000 and XC2000L which is a nominal 3.3 V version of the standard range. Each family has just two members, a 64CLB version (XC2064/L) and a 100CLB version (XC2018/L). The basic speed grades offer 10 ns pin-to-pin delays and 70 MHz clock frequencies, although faster versions of the standard devices

Table 5.1 Xilinx LCA families: summary

Part number	I/O cells	Logic cells	Flip-flops per cell	Inputs per cell	Min. T_{pd} (ns)	Icc (0 MHz) (mA)
XC2064	58	64	1	4	5.5	5
XC2018	74	100	1	4	5.5	5
XC3020/A/L	64	64	2	6	4.1(A)	0.02(L)
XC3030/A/L	80	100	2	6	4.1(A)	0.02(L)
XC3042/A/L	96	144	2	6	4.1(A)	0.02(L)
XC3064/A/L	120	224	2	6	4.1(A)	0.02(L)
XC3090/A/L	144	320	2	6	4.1(A)	0.02(L)
XC31xx/A	Configuration as XC30xx above, performance as XC3195A					
XC3195/A	176	484	2	6	2.7	5
XC4002A	64	64	2	10	4.0	10
XC4003/A	80	100	2	10	4.0	10
XC4004A	96	144	2	10	4.0	10
XC4005/A	112	196	2	10	4.0	10
XC4006	128	256	2	10	4.0	10
XC4008	144	324	2	10	4.0	10
XC4010	160	400	2	10	4.0	10
XC4013	192	576	2	10	4.0	10
XC4025	256	1024	2	10	4.0	10
XC4003H	160	100	2	10	4.0	10
XC4005H	192	196	2	10	4.0	10
XC5202	88	64	4	20	6.0	10
XC5204	112	100	4	20	6.0	10
XC5206	148	196	4	20	6.0	10
XC5210	192	324	4	20	6.0	10
XC5215	244	484	4	20	6.0	10

may be clocked at up to 130 MHz. Quiescent power is less than 12 mA for the standard range, depending on the interface, but below 20 μA for the XC2000L family.

There are five XC3000 series variants. XC3100 is a faster XC3000 with toggle rates up to 270 MHz, compared with 125 MHz for the basic family. 'A' versions of both devices have enhanced routing capability, with direct connection to the tri-state buffers on the horizontal long lines; in the standard device, this connection is made via the short lines. There are some other improvements over the XC3000/3100 families. Finally, there is the

XC3000L, architecturally identical to XC3000A, but with 3.3 V supply and 20 µA quiescent current.

Array sizes range from 64 CLBs to 484 CLBs, the largest having up to 176 user I/Os. The largest device (XC3195/A) is not available in any of the 3000 families, where 320 CLBs and 144 I/Os is the most on offer.

Three XC4000 series families are also available. Their performance is the same with the fastest of each family specified at 4 ns delay through a CLB and 10 mA quiescent current. The differences lie in architectural detail. XC4000A has less routing resources than the XC4000 and may, therefore, provide a more economical solution. On the other hand, its output pins have a higher specification in terms of drive capability and slew rate control.

The XC4000H also differs in architecture by having more I/O resources than the other two families. The XC4000 range has from 100 CLBs to 1024 CLBS with 80 to 256 I/Os; the XC4000A has 64 to 196 CLBs with 64 to 112 I/Os, but the XC4000H has, at present, just two members with 100/196 CLBs and 160/192 I/Os. This range of three families, then, caters for most combinations of logic complexity and interface requirement.

The XC5000 family has only just been announced, at the time of writing. Three devices are planned, initially, with from 196 to 484 CLBs and 148 to 244 I/Os. Performance is expected to be at about 40 MHz system clock, with 5 ns delay across a cell. The XC5000 parts are pin-compatible with XC4000 parts, in the same package, and will offer a lower cost solution where some performance compromise can be made.

The main features of the Xilinx LCA families are listed in Table 5.1.

5.1.6 Using Xilinx LCAs

LCAs are commonly designed with one of the industry-standard Schematic Capture packages, as described in Chapter 3, although an interface to ABEL-HDL is also available.

The designer must use Xilinx component libraries with the intended schematic capture tool; originally, each FPGA family had its own library, but these have been replaced by a Unified Library which covers all families. This makes it easy to migrate a design to a different family, a useful facility if the performance of a design in one family does not meet specification.

The usual design route is followed – design entry, functional simulation, layout and timing simulation.

There are two interfaces in this flow; first the schematic/HDL has to be converted to a Xilinx netlist. Then, after layout, timing data is inserted into the schematic to give an accurate prediction of actual delays. This process is called back-annotation.

The suite of programs supplied by Xilinx is called XACT (Xilinx Automatic CAE Tools). The layout is performed by PPR (Partition, Place-and-route), which may be run automatically, or with manual intervention using XDE (XACT Design Editor) to preplace CLBs or critical tracks. The result of the layout is defined by a bitstream file which is used to configure the target device. This process uses some dedicated pins to control configuration, but most of the pins have a dual function which revert to user I/O after configuration.

In the development phase the bitstream is usually downloaded via a cable connected to the computer or workstation being used for the design. In production, the bitstream is stored in a serial PROM. If more than one FPGA is used in a system, they may be daisy-chained and all the data stored in a single PROM. Xilinx supply serial PROMs dedicated to this application. It is also possible to store configuration data in standard EPROMs and load one FPGA in parallel; successive devices must be daisy-chained in serial mode, however.

Full details of configuration and start-up can be supplied by Xilinx.

5.2 Atmel FPGAs

5.2.1 AT6000 array

The AT6000 series is the only family marketed by Atmel, at present. Like the Xilinx families it is a RAM-based FPGA, needing to be configured whenever it is powered up. The Atmel array is based on an 8×8 block of cells, repeated across the FPGA. Each chip is a square matrix of blocks; the smallest having four on each side, the largest ten. The cells are not as complex as the Xilinx CLBs, but have fast interconnections within the block; this allows high performance macros to be built from adjoining cells.

A group of four cells, at the intersection of four blocks, is shown in Figure 5.10. Each cell has two inputs, two outputs and an I/O on every side. The inputs and outputs are potentially connected to an adjacent cell, while the I/O is connected to the adjacent local bus. There is also an express bus which may connect with a local bus at one of the repeaters at the edge of the block. The repeaters are uni-directional, except for two of the local busses in each cell. These have an augmented connection for bi-directional transfer, and may be used to implement tri-state busses.

The busses can only transmit signals in one direction 'North-South' or 'East-West'. Each cell, however, contains a programmable connection between two local busses; this connection may be established without interfering with the cell's logic function. Signals may, therefore, turn a corner at any cell site.

RAM-based FPGAs   117

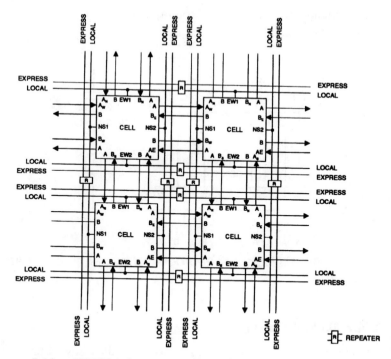

Figure 5.10 *AT6000 interconnections (reproduced by permission of Atmel Corporation)*

For general logic paths, then, there are three levels of interconnect to allow a hierarchical structure to be built up. Simple macros may be established by using direct cell-to-cell connection. These may be combined into more complex functions within a block by using the local bus. Signals needing to cross block boundaries can use the express bus. Because this only has connecting points at the repeater sites it will suffer little speed degradation, compared with the local bus which has eight connections in each block.

Apart from the logic connections, there is a separate distribution network for the clock and reset to the flip-flop in each cell. Each column has an associated clock and reset line with one of four sources for each. The four clock options are the global clock, the express bus at the head of the column, an output from the cell at the head of the column or Vcc. Similarly, the reset may be sourced from the global reset, the express bus at the foot of the column, an output from the cell at the foot of the column or Vcc. There is, thus, some flexibility in the clock sourcing but, because each column must share the same clock, true asynchronous logic may be difficult to implement. In some circumstances, this could be a blessing in disguise.

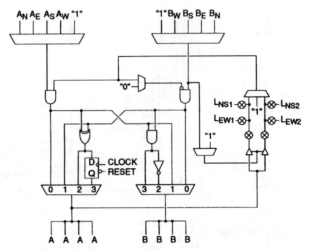

Figure 5.11 AT6000 logic cell (reproduced by permission of Atmel Corporation)

5.2.2 AT6000 logic and I/O cells

The AT6000 logic cell, Figure 5.11, is simpler than any Xilinx CLB. Although RAM-based FPGAs tend to be coarse grained, because of the relative slowness of the interconnects, the direct cell-to-cell connections of the Atmel array permits a fine grained structure to deliver high performance.

As we saw in the previous section, each cell has two inputs, two outputs and an I/O on each side, in a symmetrical layout. The two sets of respective inputs from the four sides are each multiplexed together with a fixed HIGH, so that the logic functions can be formed with a signal entering from any side of the cell. One of the local bus signals from any side may also be selected, in a third multiplexer, to provide a third input to the cell logic.

Although the cell contains only four gates, an invertor and a flip-flop, there are 44 possible configurations, ranging from an invertor to an exclusive-OR driven flip-flop, which can form the basis of a counter stage. The flip-flop, which may be by-passed, drives one output on each side of the cell and may be routed to one or more local bus connections. These have an optional tri-state buffer.

Each cell also has a second, combinatorial output. This, again, is routed to all four sides of the cell to make signal propagation independent of direction. The combinatorial output cannot be connected to the local bus; it is only available for connection to adjacent cells.

The logic functions available from the primitive cells are, because of the simplicity of the cell, fairly basic. By making direct connections available to the cells on all four sides it is possible to define hard macros of almost any complexity. Because the direct connections contain only a single switch, they are fast compared with the usual programmable interconnect found in RAM-based FPGAs. The hard macros will, therefore, have a fast dynamic performance. Because the cells are perfectly symmetrical, any hard macro can be transformed by reflection along the NS or EW direction, making layout very flexible.

Each pad in an AT6000 FPGA is associated with an 'entry cell' and an 'exit cell'. These form the interface between device pins and internal logic. Alternate pads connect to the 'A' inputs and outputs, and to the 'B' inputs and outputs. The usual features, such as TTL/CMOS compatibility and slew rate control, are available in the buffers which connect the pads to the cells.

Unlike core logic cells, the I/O cells have connections to the express bus, inputs in exit cells and outputs in entry cells. I/O signals thus have immediate access to the interior of the FPGA.

The tri-state control for the output buffer may be 'OFF' (the default condition), 'ON' or controlled by one of two local busses associated with the exit cell. Because the I/O signals are connected directly to logic cells, any input or output registering may be performed by the flip-flop in the respective entry or exit cell.

5.2.3 Design and use of Atmel FPGAs

Atmel FPGAs are designed in the same overall manner as most other FPGAs. A schematic or VHDL entry system interfaces to the Atmel layout system via a library of macros. As mentioned above, hard macros are available to connect cells in a predictable manner and generate high-level functions. In many cases the hard macros have been optimized for either performance or cell usage. Thus, the designer can decide which is more important for his application. Soft macros may also be used for extra flexibility at the layout stage.

After functional simulation, the program extracts a netlist and proceeds to the place-and-route stage. Successful layout may be followed by back-annotation of timing parameters to the schematic, allowing the designer to check predicted delays in his circuit. Small changes may be made without disturbing existing structure. As with Xilinx devices, the end result of a design is a bit-stream which is used to configure the target device in-circuit.

In production, the bit-stream data may be contained in an EPROM, in serial or parallel format, a microprocessor or downloaded from a PC port. One important feature of Atmel configuration, which is not supported by all manufacturers, is configuration 'on the fly'. All or part of an AT6000 device may be reconfigured during normal operation. The term 'Cache

Table 5.2 AT6000 family summary

Part number	I/O cells	Logic cells	Flip-flops per cell	Inputs per cell	Min. T_{pd} (ns)	Icc (0 MHz) (mA)
AT6002	96	1024	1	3	2.2	0.5
AT6003	120	1600	1	3	2.2	0.5
AT6005	108	3136	1	3	2.2	0.5
AT6010	173	6400	1	3	2.2	0.5

Logic' has been coined for this mode of use, by association with cache memory.

In many cases, some of the functions in an FPGA are not needed for the whole operating time of a system. 'Cache Logic' means that only those parts of an FPGA being used need be configured at any one time. If some of the functions are mutually exclusive, they can be loaded into the same part of the chip as and when needed. A modem, for example, may need a serial-to-parallel convertor while receiving, but parallel-to-serial when sending. These functions would usually both be provided in a permanent configuration; with Cache Logic, however, the same area of the FPGA could be programmed with either function according to the operating mode, meaning that a smaller, cheaper device could be used.

Because the partial reconfiguration can take place while the rest of the circuit is operating normally, any other functions in the FPGA are unaffected.

The current available devices in the AT6000 family are listed, with their chief features, in Table 5.2.

5.3 FLEX8000/10K families

5.3.1 Logic array block structure

Altera's FLEX8000 and 10K devices are based on LABs (logic array blocks) in a coarse-grained structure, surrounded by FastTrack row and column interconnects. Each LAB contains eight simple LEs (logic elements) and a 32-channel local routing matrix. Figure 5.12 shows the LE layout and Figure 5.13 shows how they fit together with the local interconnect to form an LAB.

Each LE has ten inputs and a single output. Four inputs are derived from the local bus, four are control inputs, common to each LE in the LAB, and

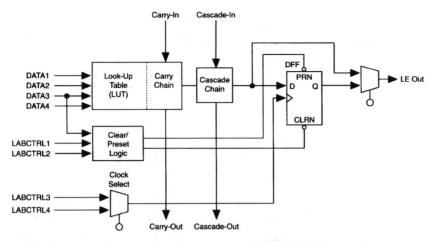

Figure 5.12 *FLEX logic element (reproduced by permission of Altera Corporation)*

two are a carry-in and cascade-in. The four data inputs feed an LUT (look-up table) for implementing a wide range of combinatorial logic functions; the resulting signal may be output as a registered or direct function. The control inputs provide a choice of clocks, and a clear and preset to the flip-flop. They may be sourced from dedicated FPGA inputs or from the local interconnect.

10K parts have an additional logic block associated with each row of LABs. This is the EAB (embedded array block) which is, essentially, an eight-input eight-output RAM. It is intended to implement 'megafunctions', such as multipliers or simple processors, as well as memory. These functions can be created more efficiently, in terms of LE and interconnect usage, from a look-up table than from the resources in a standard LAB.

The LAB has a carry and cascade chain running through it, and through adjacent LABs in the same row. The carry chain may be used for building fast counters and adders. In arithmetic or counter mode the LUT is split into two three-input LUTs. One generates the normal logic output for the stage and uses the carry-in and, in counters, register feedback as two of the inputs; the other small LUT generates carry-out for direct connection to the next stage.

The cascade chain takes the normal logic output from the LUT directly into the next stage. This enables wide logic functions to be built with a significantly reduced delay penalty, compared with cascading through the local interconnect. In addition, the 10K family has a tri-state chain with programmable pull-up or pull-down. Two of the LUT

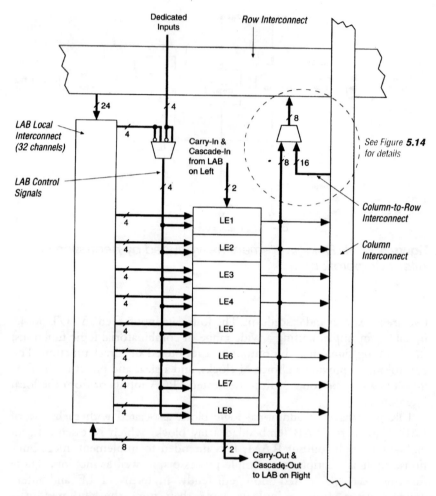

Figure 5.13 *FLEX LAB layout (reproduced by permission of Altera Corporation)*

inputs act as data lines, the other two act as output enables; the tri-state line may be connected to the normal routing channel via the LE output line.

These additional connections make it possible to implement common functions efficiently without undue use of general routing resources. As with AT6000 arrays, the direct connections make the implementation of high speed macros possible. Examples are a sixteen-bit counter operating at up to 95 MHz, using two LABs, or a 16-bit decoder in half an LAB with a delay of under 5 ns.

5.3.2 FLEX interconnect and I/O resources

The LABs are set in a matrix of horizontal and vertical routing channels. Each horizontal channel is associated with a row of LABs, with the ability to route signals into or out of each LAB. There are, typically, 21 LABs and 16 I/O elements associated with each row which have, in most cases, 168 routing channels.

The vertical columns run between the LABs and contain 16 channels each. Each LE output can drive up to two channels but, because there are only between two and six rows over the whole FLEX8000 range and from three to thirteen rows in the FLEX10K range, most of the connections will be in the horizontal direction. There are also four I/O elements associated with each column.

At each crossing point there are eight three-input multiplexers which can route one of two vertical signals or an LE output to a horizontal channel. There is no way of routing a horizontal signal to a vertical channel, without passing through an LE.

The eight I/O elements at either end of the rows are driven by a wide multiplexer, arranged so that each column has a potential connection to each device output. Each I/O can drive up to two row channels when used as an input. Similarly, each column I/O has an eight-way multiplexer driving it, enabling any LE in the column to be connected to any I/O.

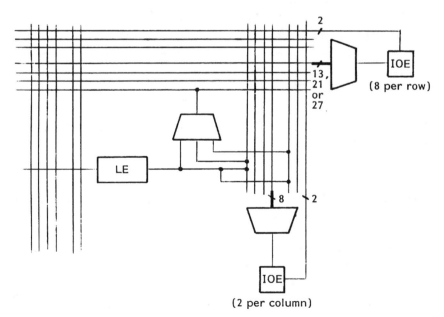

Figure 5.14 *FLEX interconnection scheme*

When used as an input, an I/O element can drive two column channels. Figure 5.14 shows one corner of a FLEX device with details of row/column cross-overs and I/O element connections.

The I/O elements themselves contain a flip-flop for registering either the input or output signal, a programmable invertor and an output buffer with tri-state capability and slew rate control. The control signals for the I/O elements are distributed round the chip on a peripheral bus. There are four output enables, and two clocks and two clears for the I/O flip-flops. They are derived from either the four dedicated control inputs or from the internal logic.

Because every I/O signal traverses the whole chip in either the horizontal or vertical direction, the allocation of signals to I/O can be quite flexible. This is an advantage when redesigning chips where the PCB layout is already defined; it also makes the performance predictable, as all connections in each direction are the same length.

5.3.3 FLEX8000/10K design, performance and range

The MAX+PLUS II development system supports mixed schematic, waveform and text entry design methods, including Altera HDL, VHDL and Verilog HDL. It also provides a netlist interface to the usual industry standard design tools. The design flow is, as usual, logic entry, functional

Table 5.3 *FLEX families: summary*

Part number	I/O cells	Logic cells	EAB cells	Flip-flops per cell	Inputs per cell	Min. T_{pd} (ns)	Icc (0 MHz) (mA)
EPF8282(A)(V)	78	208	0	1	4	2.4	0.5
EPF8452(A)	120	336	0	1	4	2.4	0.5
EPF8636A	136	504	0	1	4	2.4	0.5
EPF8820(A)	152	672	0	1	4	2.4	0.5
EPF81188(A)	184	1008	0	1	4	2.4	0.5
EPF81500(A)	208	1296	0	1	4	2.4	0.5
EPF10K10	148	576	3	1	4	2.4	0.5
EPF10K20	196	1152	6	1	4	2.4	0.5
EPF10K30	244	1728	6	1	4	2.4	0.5
EPF10K40	276	2304	9	1	4	2.4	0.5
EPF10K50	308	2880	10	1	4	2.4	0.5
EPF10K70	356	3744	9	1	4	2.4	0.5
EPF10K100	420	5408	13	1	4	2.4	0.5

simulation, place-and-route, and post-layout simulation with timing analysis.

The development system also provides a direct interface to a download cable which may be used to configure FPGAs in-circuit, or to program configuration EPROMs. These hold the bitstream data, and may configure a FLEX device in parallel or serial mode. Alternatively, the devices may be configured in-circuit with a local intelligent host, in which case the bitstream may be derived from a mass storage device, such as a hard disk, saving board space by omitting the EPROM. However the FPGA is configured, seven or eight pins are dedicated to the configuration process and are not available as user I/O.

FLEX devices are available in several speed options, 5 V and 3.3 V interfaces, and packages from 84-lead PLCC to 560-pin PGA. Standby current is, typically, less than 1 mA for both power voltage choices and, as we saw above, an up/down counter will operate at up to 95 MHz.

The quoted 'usable' gate counts vary from 2500 to 16 000 across the FLEX8000 range of six device types, and from 10 000 to 100 000 gates for the seven FLEX10K devices. Table 5.3 lists the whole range.

5.4 Other RAM-based FPGA families

5.4.1 ORCA 2C series

Optimized reconfigurable cell arrays (ORCA) are manufactured by AT&T Microelectronics. They are characterized by high cell and I/O counts and use a coarse-grained structure with variable length routing busses to achieve a high logic packing density.

The basic logic cell is called a PFU (programmable function unit), see Figure 5.15, and contains look-up tables (LUTs) for combinatorial logic and four flip-flops which can register the LUT functions. The PFU has sixteen inputs, twelve of which can be combined in various LUT modes, and four which may drive one of the flip-flops directly. Cells near the edge of the chip can also accept inputs directly from the I/O cells into the flip-flops. There are six outputs from each PFU; five are selected from the four LUT outputs and four flip-flops, the sixth is a carry-out.

The LUT can act as four separate four-input tables, two five-input tables or as two 16 × 2 RAMs. It can also mix modes so that a single PFU contains a 16 × 2 RAM and a five-input logic function. In four-input mode, five inputs are shared between the two blocks in each half of the PFU. In five-input mode, the two outputs can be multiplexed or exclusive-ORed together, under control of an eleventh input, C0. The twelfth input is a carry-in, which is used to cascade arithmetic or counter stages by direct connection of the carry-out with neighbouring PFUs.

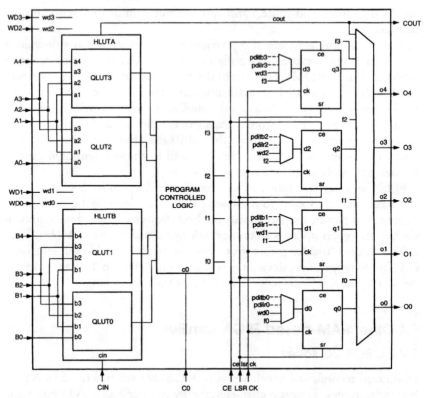

Figure 5.15 *ORCA PFU cell (reproduced by permission of AT&T Microelectronics)*

In RAM mode, the five 'logic' inputs to each half of the LUT act as address and write enable, data is input on two of the lines which drive the flip-flops directly, and output on the usual LUT outputs.

There are three control inputs to the flip-flops, clock, clock enable and set or reset which can be programmed individually for each flip-flop. There is also a global set/reset line which operates independently from the local signal; if the local set/reset is not used it can control a multiplexer to select the source of data input to the flip-flop. The clock enable also controls a multiplexer, feeding either the present state or new data to the flip-flop. The flip-flops themselves may be programmed as positive or negative triggered flip-flops or latches.

With this level of sophistication in the PFU, equivalent to an MSI function in a standard logic family, it is clear that quite extensive routing resources are required. Each cell has eight short lines (one cell span) and eight long lines (four cells span) associated with it in the horizontal and vertical directions. There are also four lines in each direction, spanning half the array, and four spanning the whole array. The shorter lines can be extended by enabling programmable connections with adjacent lines.

Signals into or out of each PFU are routed via switching nodes, which are located at the corners of each PFU. There are also bi-directional tri-state drivers associated with each PFU for creating internal tri-state busses on the long lines, or for driving signals on the long lines to minimize routing delays.

The programmable I/O cells (PICs) are located around the periphery of the chip and each has four pads and four I/O buffers associated with it. The buffers offer the usual features of FPGA I/Os, such as pull-up and pull-down, TTL/CMOS compatibility, programmable slew rate and programmable delay, with a tri-state option on the output line.

The I/O signal lines are connected to a switching matrix which routes them to the various lines spanning the chip, as described above. It can also connect them directly to adjacent PFUs for fast access, or for providing input or output registering. Signals are, additionally, routed around the periphery of the chip on a bus structure similar to the busses interconnecting the PFUs. Connection to the PFU bus system takes place in the PIC switching matrix.

The recommended design tool for ORCA FPGAs is the NeoCAD FPGA Foundry, although other standard design capture tools may be used for the front end. The outcome, as with other RAM-based FPGAs, is a bitstream which is used to configure the switches inside the ORCA chip. Serial or parallel downloading may be used for configuration; AT&T make a series of serial ROMs which can be used for holding the bitstream, or standard EPROMs used in parallel mode.

The ORCA range extends from a 100 PFU device to 576 PFUs in a 429-pin pin grid array. Details are listed in Table 5.4.

Table 5.4 ORCA 2C series FPGAs

Part number	I/O cells	Logic cells	Flip-flops per cell	Inputs per cell (ns)	Min. T_{pd} (mA)	Icc (0 MHz)
2C04	160	100	4	16	3.9	1.5
2C06	192	144	4	16	3.9	2.0
2C08	224	196	4	16	3.9	2.7
2C10	256	256	4	16	3.9	3.4
2C12	288	316	4	16	3.9	4.2
2C15	320	400	4	16	3.9	5.0
2C26	384	576	4	16	3.9	7.2

5.4.2 MPA1000 FPGAs

As a complete contrast to the ORCA series, the final family we shall examine in this chapter is a fine-grained FPGA from Motorola. The logic cells are laid down in a four-level hierarchy, allowing routing based on timing requirements across the chip.

Figure 5.16 shows the basic cell and the way it fits into the hierarchy. Each cell contains a two-input NAND gate and a secondary function; this function depends on its position in the next level, the tile. Tiles have four primary cells in a 2 × 2 arrangement; their secondary functions are two exclusive-ORs, a wired-OR and a flip-flop/latch. The XORs have a fast carry chain propagated through them for counter/arithmetic functions while the flip-flop cell has a clock and reset as direct inputs. A multiplexer at the output selects either the NAND function or secondary function.

The lowest level of routing is a direct connection from the output to both input multiplexers of the two cells on the immediate left and right. It also connects to one input multiplexer of the cells to the left and right of these and to the two cells above and below. Each output also drives up to four medium busses which traverse the next hierarchical level, the zone.

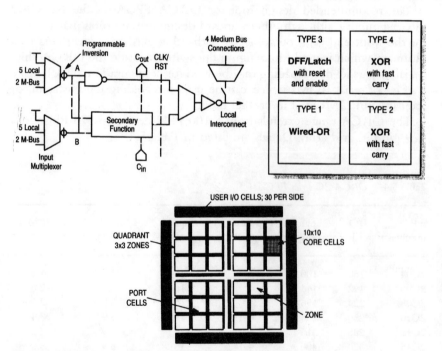

Figure 5.16 *MPA1000 cell hierarchy (reproduced by permission of Motorola Inc.)*

There are two medium bus inputs to each input multiplexer, one horizontal and one vertical.

A zone contains 100 core cells, that is 25 tiles. Around the edge of each zone are port cells which connect all levels of bus, except the local interconnect. The chip is divided into four quadrants, each of which has a square array of zones, 3×3 in the smallest to 5×5 in the largest of the three family members. Each quadrant has a set of global busses which run the length and breadth of the quadrant.

Global busses can only connect to lower levels of hierarchy at the port cells at the zone edges. Four medium bus lines enter each port cell, two can connect to the global bus while the other two can join medium bus lines from the neighbouring zone. There is also an X-bus traversing the chip; this has a switch at every crossing to allow right-angle turns for long signal paths. It can also be used to connect medium bus lines across zones, although the global bus is faster. Right angle turns in the medium bus can be made at any logic cell which does not have an output connected directly to the medium bus.

The clock and reset lines are also distributed at the port cell, either from the dedicated low skew clock bus, or from a global bus. Eight different input sites can be used as the source of the dedicated clock bus.

The I/O cells are distributed around the chip periphery at the rate of one per tile; they have built-in flip-flops for both input and output and the usual programmable features – slew rate, drive, TTL/CMOS compatibility and 3 V/5 V level. They may be connected directly to the routing array or to a peripheral bus for distribution around the chip.

Design input is by one of the usual commercial entry packages and NeoCAD is also used for layout and timing simulation. NeoCAD also generates the bitstream which may be programmed into a serial EPROM for device configuration.

There are, at present, three family members whose characteristics are summarized in Table 5.5.

Table 5.5 *MPA1000 series FPGAs*

Part number	I/O cells	Logic cells	Flip-flops per cell	Inputs per cell (ns)	Min. T_{pd} (mA)	Icc (0 MHz)
MPA1036	120	3600	0.25	2	2.81	tba
MPA1064	160	6400	0.25	2	2.81	tba
MPA1100	200	10000	0.25	2	2.81	tba

Note 1. Estimated delay for a 3-level NAND function.

6 Antifuse FPGAs

6.1 Actel families

6.1.1 ACT1 structure

ACT1 is the name of the first antifuse FPGA family introduced by Actel, who use the silicon/silicon structure with an oxide/nitride barrier providing the insulation before fusing.

Physically, ACT1 devices consist of rows of logic modules separated by horizontal routing channels, each with 22 tracks. There are also vertical routing tracks, with thirteen tracks per logic module. The tracks are segmented into different lengths to give the option of making short connections without using long tracks. Short segments may be connected by an antifuse to make longer connections; there are also programmable connections at the crossing points between horizontal and vertical tracks. Thus, a good mix of track lengths is available for efficient and comprehensive routing.

The logic module, whose circuit is shown in Figure 6.1, has eight inputs and one output, and can implement most basic logic functions of up to four inputs, plus a selection of more complex functions such as latches, multiplexers and exclusivity functions. The functions are not configured by programming internal cells, as with RAM-based FPGAs, but by connecting the appropriate module inputs to Vcc or ground. Figure 6.2 illustrates this for a three-input AND gate.

Only when the select inputs for multiplexers M2 and M3 are HIGH will the signal on input 'A' be transmitted to the output; thus all three inputs must be HIGH if the output is to be HIGH. Figure 6.3 shows the effect of repositioning the inputs; now the output will always be HIGH unless the inputs are LOW. We now have an AND gate with inverting inputs, or a NOR gate.

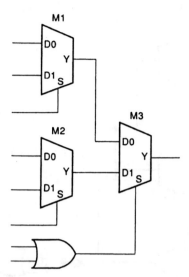

Figure 6.1 *ACT1 logic module (reproduced by permission of Actel Corporation)*

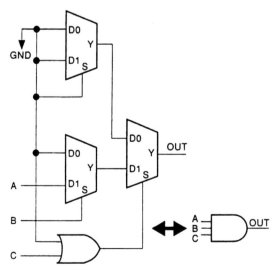

Figure 6.2 *ACT1 module configured as AND gate (reproduced by permission of Actel Corporation)*

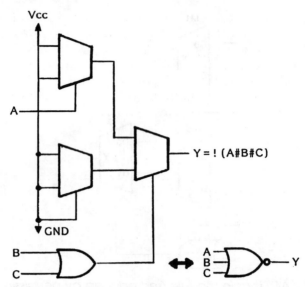

Fig 6.3 ACT1 module configured as NOR gate (reproduced by permission of Actel Corporation)

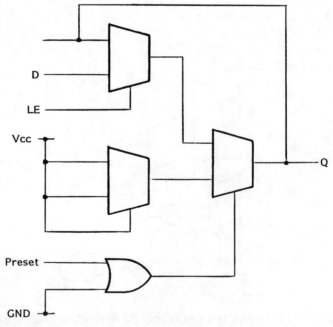

Figure 6.4 ACT1 module configured as a D-latch

It is also possible to build more complex functions, such as latches and complex gates. In Figure 6.4 an active-HIGH latch with active-HIGH preset is illustrated. The upper multiplexer implements the basic latch function. While preset is LOW the lower multiplexer is locked out, but when it goes HIGH the output is taken HIGH irrespective of the levels of D and LE. A D-type flip-flop can be built from two latches in master-slave configuration.

There is a high fan-out clock network allowing one common signal to be fed to a large number of flip-flops with low skew.

In the physical FPGA layout, each logic module input or output connects to a horizontal channel via a vertical track; inputs are tied to Vcc, ground or a horizontal track according to which function is being implemented in the module.

The I/O buffers are quite simple, offering input, output, bidirectional or tri-state capability with CMOS and TTL compatibility. Any input or output registering must be done in a logic cell.

6.1.2 ACT2 series

The ACT2 FPGAs use two types of logic cell, a combinatorial or C-module and a sequential or S-module.

The C-module, shown in Figure 6.5, is an enhanced version of the ACT1 logic cell. It is a four-input multiplexer with gated select inputs, one select via an OR gate the other via an AND gate. This allows higher fan-in gates, such as a five-input AND gate, to be included as a single-cell macro. Several of the ACT1 four-input gates must be built from two logic cells but, with the increased logic content of the C-module, very few ACT2 gates need two modules.

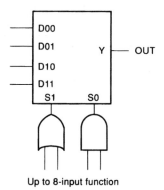

Figure 6.5 *ACT2 C-module (reproduced by permission of Actel Corporation)*

The S-module is virtually a C-module driving a flip-flop, except that one of the AND gate select inputs is used as a reset. Alternative S-module configurations are a latch replacing the flip-flop, or the same function as a C-module.

As with the ACT1 series, there is a horizontal routing channel between the module rows containing segmented tracks which may be connected end to end by antifuses.

Vertical tracks are either connected to module inputs and outputs, or they are uncommitted and segmented, like the horizontal tracks. Most cross-overs between horizontal and vertical tracks have a potential antifuse connection. Input vertical tracks span only one row, either above or below the module; output vertical tracks extend for two rows in either direction. Module outputs can also be connected directly to the uncommitted vertical tracks; in this case the antifuse is programmed at a high current level to reduce series resistance.

The modules are laid out in pairs along the rows, two C-modules next to two S-modules, and so on. There are I/O modules at each end of the rows, and in two rows at the top and bottom of the chip. The row I/Os can accept signals from either horizontal channel, above or below the row, for output from the I/O pad, while the top and bottom I/Os have a single

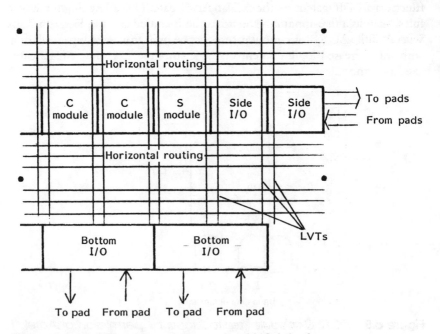

Figure 6.6 ACT2 routing scheme and I/O interface

output source. Inputs and outputs can be latched in the I/O buffers, which are tri-statable and TTL/CMOS compatible.

Side I/Os (at the ends of rows) have dedicated vertical tracks, as logic modules, but I/Os at the top and bottom of the array connect to non-dedicated vertical tracks via an antifuse. There may, therefore, be some timing differences between the side and top/bottom I/O modules. Detail of the way in which I/Os interface to the routing scheme is shown in Figure 6.6.

There are two dedicated clock networks in the ACT2 family, located in each horizontal channel with drivers at the input to each channel. The clocks may be sourced from two external clock pins or from two internal sources. There are fewer antifuse connections between the clock tracks and logic modules, than the usual routing tracks; this reduces capacitative loading on the clock networks.

The improvements in layout, compared with ACT1, give the ACT2 family a performance edge which is illustrated by the increase in clock frequency from 65 MHz to 130 MHz, for the smallest array in each family.

6.1.3 ACT3 series

ACT3 FPGAs are also based on a dual module layout. The C-module is functionally equivalent to the ACT2 series C-module. The S-module is a full C-module driving a flip-flop with clock and clear, as in Figure 6.7. As with ACT2, the flip-flop may also be configured as a level-sensitive latch.

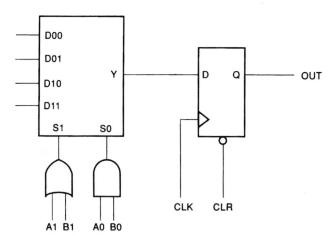

Figure 6.7 *ACT3 S-module (reproduced by permission of Actel Corporation)*

The standard interconnect scheme is practically identical to that described for ACT2, with the exception that there is an extra vertical track per module pair.

The I/O modules are an enhanced version of the ACT2 cells; there are built-in flip-flops with dedicated I/O clock and set/reset lines around the whole chip. One innovative feature is that the registered output can be fed back through the input lines; this allows the output register to be used for internal logic, as well as its usual function. The side I/O cells may connect to the channel above or below the module row, similarly to ACT2, while the top and bottom modules have to be routed to the uncommitted vertical tracks.

The clock network is an extended version of the ACT2 clocks. As well as the general-purpose routed clocks, there is a high-speed hard-wired clock (HCLK) which contains no antifuse elements between the driver and module input. It delivers a guaranteed performance and very low skew. The routed clocks may be used for other high fan-out signals, such as resets or enables.

Performance should be extended to a 150 MHz clock frequency by using HCLK in the ACT3 family.

6.1.4 Using Actel FPGAs

The mechanics of designing Actel FPGAs follows the usual path. Design entry uses one of the standard schematic capture packages, with options of equation or VHDL blocks. The logic is proved by simulating the circuit and then an interface program converts the captured logic blocks into a netlist capable of being recognized by the layout software. After design rule checks and optional pin definitions the process continues with place-and-route. Chip performance can now be predicted by extracting the timing information and re-simulating with estimated delays in each circuit net.

Once the device performance has been optimized, a programming file is produced. This specifies which antifuses are to be blown to produce the required circuit configuration. At the horizontal and vertical antifuse sites there are also pass transistors which are used to select the cross antifuses to be programmed. As programming proceeds across the chip, more connections are made between horizontal and vertical lines, so the order in which antifuses are programmed becomes significant. Moreover, it is not usually possible to go back and reprogram fuses if, for example, they failed to program at the first attempt, or if a change is required to an existing pattern.

The actual programming sequence involves pre-charging all the device segments to half the programming voltage (Vpp). Any unaddressed fuses will then experience only one half Vpp across them even if one side is at ground or full Vpp. Fusing is detected by current flowing in the fusing circuit; further pulses of current are then applied to burn the fuse in.

The pass transistors are also used to test unprogrammed devices. While reprogrammable parts can be programmed, tested and then erased, it is not usually possible to test one-time programmable devices before they are programmed. This can lead to a relatively high functional failure rate in PLDs. Actel FPGAs are 100% testable before programming. A shift register round the chip periphery can be loaded with data to turn on selected pass transistors and apply test patterns. The same register is used to select antifuse locations during the programming phase.

After programming is complete there is a useful diagnostic tool called the Actionprobe. This connects to the programmer and to the programmed device, which may be in-circuit, and sends the address of internal nodes via dual-purpose pins SDI and DCLK. The state of the addressed nodes can be examined on two other dual-purpose pins, PRA and PRB. These pins are set to diagnostic mode by taking the MODE pin HIGH.

Because the diagnostic pins may share functionality with I/Os, these I/Os must clearly be non-critical functions if probing is to be used. The probe function is disabled by setting the security fuses; if the SDI and DCLK pins are to be used as outputs the security fuses must be set. If probing is not required, and the design is fixed, it is advisable to set the

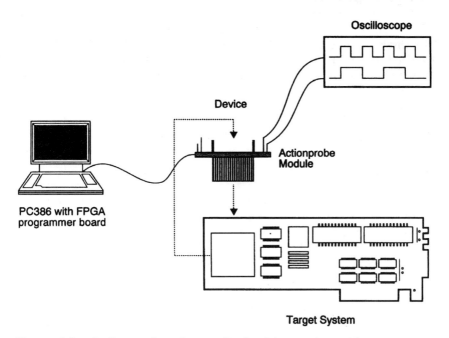

Figure 6.8 *Actionprobe diagnostic tool (reproduced by permission of Actel Corporation)*

security fuses to prevent unwanted probing by third parties; it also prevents further programming – a useful safety feature.

Figure 6.8 illustrates the Actionprobe diagnostic tool.

6.1.5 Actel FPGA performance and range

Actel FPGAs are, essentially, low-power devices at zero frequency with 2–3 mA maximum supply current. As with any CMOS device, the dynamic power depends on the frequency at which the various internal nodes are switching, and the loading on the outputs. In a typical application with about 2000 gates operating with a 20 MHz system clock the supply current will be of the order of 100 mA.

Delay times are dependent on the fan-out from each logic module. In the ACT1 standard speed grade this can vary from 5.9 ns for fan-out of one to 14.7 ns for eight internal loads. This is partly due to the extra loading of the antifuses needed to connect the extra loads, and partly because long lines are almost certainly needed for high fan-out circuits. There is an advisory limit of ten loads; nets designated as critical are limited to a fan-out of four.

In the speed column of Table 5.3 we give the figure for a fan-out of four in the best speed grade available.

There are just two parts in the ACT1 range, but they are available in a variety of packages, according to the number of user I/Os required. ACT2 has three family members with a maximum of 140 I/Os and 8000 gates claimed; the ACT3 family, of five parts, extends to 10 000 gates with 228 I/Os. They are all listed in Table 6.1.

Table 6.1 *Actel FPGA families*

Part number	I/O cells	Logic cells	Flip-flops per cell	Inputs per cell	Min. T_{pd} (ns)	Icc (0 MHz) (mA)
A1010	57	295	0.5	8	7.0	3
A1020	69	547	0.5	8	7.0	3
A1225	83	451	1/0.5[1]	8	6.6	2
A1240	104	684	1/0.5[1]	8	6.6	2
A1280	140	1232	1/0.5[1]	8	6.6	2
A1415	80	200	1/0.5[1]	8	4.8	2
A1425	100	310	1/0.5[1]	8	4.8	2
A1440	140	568	1/0.5[1]	8	4.8	2
A1460	168	768	1/0.5[1]	8	4.8	2
A14100	228	1153	1/0.5[1]	8	4.8	2

Note 1. One flip-flop per S-module, one flip-flop per two C-modules

6.2 QuickLogic/pASIC380 FPGAs

6.2.1 QL technology and structure

QuickLogic have introduced the QL family of FPGAs, which is also available from Cypress Semiconductors. It uses a metal/metal antifuse which is smaller than the silicon/silicon structure employed by Actel; because of this it has a lower ON resistance and shunt capacitance. The physical size also makes it possible to fit more routing tracks into a given area.

Although its routing is probably the most compact of any FPGA technology, the QL family has a more complex logic cell than Actel and some RAM-based parts. Figure 6.9 shows that it contains six AND gates, three multiplexers and a flip-flop. Two of the gates have six inputs with an assortment of active-HIGH and active-LOW buffers, while the other four have one active-HIGH and one active-LOW input each.

The four small gates drive two two-input multiplexers, the select input coming from a six-input gate. The multiplexers feed the third multiplexer whose select input comes from the other six-input gate; this final output drives the flip-flop which has clock preset and clear inputs brought directly into the cell.

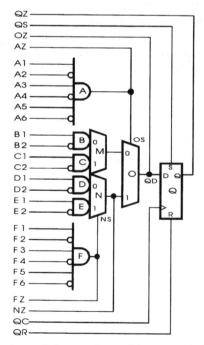

Figure 6.9 *QL logic cell (reproduced by permission of QuickLogic Corporation)*

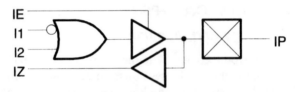

Figure 6.10 QL I/O cell (reproduced by permission of QuickLogic Corporation)

The logic cell has five outputs – each of the six-input gates, a first-stage multiplexer, the second-stage multiplexer and the flip-flop. This arrangement makes it possible to build wide gates, with or without registering, or combine two or three smaller functions in a single cell. For example, if a cell is programmed as a single input flip-flop the two wide gating functions are still available.

As with the Actel FPGAs, cell function is defined by tying unused inputs to Vcc or ground. The vertical routing channel adjacent to each cell has two lines dedicated to supplies, two lines dedicated to high fan-out clock/reset signals, and about twenty lines for logic signals. The clock/reset lines can connect only to the three flip-flop control inputs through ViaLink sites. The remaining inputs and outputs from the cell can be joined to any of the signal lines.

Horizontal routing channels run across the chips between the rows of logic cells. ViaLink elements at the crossing points between vertical and horizontal tracks allow signals to be routed transversely across the chip. Tracks may be either segmented or express. Segmented tracks have Pass Links at regular intervals and are generally used for short signal paths. Express tracks run the length of the routing array, and can carry signals over long distances with minimal speed penalty.

Some signal tracks terminate directly in I/O cells at the edge of the chip. The basic I/O cell, Figure 6.10, contains an OR gate driving an output buffer, with tri-state and open drain capability, and an input buffer. There are some dedicated input cells with high drive true/complement buffers and two input/clock cells which provide the signals to the clock lines described above.

6.2.2 QL cell characteristics

The QL cell architecture gives the pASIC designer some benefits which are not available with many other FPGAs, in particular, those with only 4–6 inputs per cell. The basic library includes gates with up to six inputs and any combination of inverting and non-inverting inputs, plus a 14-input gate with seven of each polarity. By setting appropriate inputs to Vcc or ground, many gating combinations of up to fourteen inputs are possible, with only a single level of logic delay.

Other versatile combinatorial macros which can be included in a single cell are 4-input multiplexers, with any combination of input polarities, 2 and 3-input exclusive-OR/NOR gates and 2-to-4 line decoders. Innumerable combinations of AND-OR type combination gates are possible with this cell; these are not available as hard macros, because of the number of possibilities, but are generated as needed by the logic optimizer during the assembly process.

Possible sequential macros include D-type, T-type and J-K flip-flops with enables, presets and clears, and D-latches with direct and multiplexed inputs.

The soft macros give an indication of the logic capability of the QL cell. A full adder can be implemented in two cells and may be cascaded into a 16-bit ripple sum adder occupying 32 cells. Faster adders, or accumulators, use conditional-sum architecture and, for example, a 32-bit design uses 120 cells. Because two AND gates and a 3-input XOR gate fit into one cell, a 4 × 4 multiplier requires only thirty logic cells with a five-logic level delay.

The depth of the logic cell means that functions may be generated in series within one cell. The multiplexers can be configured as two D-latches, that is a master-slave flip-flop. By combining this with the built-in flip-flop, very efficient registers can be built; an 8-bit shift register occupies just four logic cells.

As with adders, counters may be configured for speed or for density. A 16-bit ripple counter uses only 16 cells, but has a 16-logic level delay, while a 16-bit fast up counter with a single logic level delay takes just 24 cells.

By packing so much logic capability into the basic logic element, QuickLogic have produced an array which offers high performance together with efficient packing density.

6.2.3 QL design and performance

QuickLogic provide the SPDE place-and-route software to interface with standard design entry and simulation tools. The Cypress version is called Warp3; this is a self-contained design system based on VHDL, although schematic entry is possible. This may be either in Warp3 itself or as an EDIF file from a standard schematic capture program.

Both systems will place-and-route a successfully simulated design to generate a netlist and programming file. Timing parameters can then be fed back to the simulation to provide a full dynamic analysis of the final layout. Simulation can be performed graphically or with VHDL language output.

Table 6.2 QL/pASIC380 family

Part number	I/O cells	Logic cells	Flip-flops per cell	Inputs per cell	Min. T_{pd}	Icc (0 MHz)
QL8X12B/7C381A–2A	64	96	11	14	1.7	2
QL12X16B/7C383A–4A	88	192	11	14	1.7	2
QL16X24B/7C385A–6A	122	384	11	14	1.7	2
QL24X32B/7C387A–8A	180	768	11	14	1.7	2

Note 1. A second flip-flop per cell may be constructed from the combinatorial components

We have already noted that the metal/metal antifuse produces connections of lower resistance than the Actel silicon/silicon structure. QL contacts are typically $50\,\Omega$ compared with 250 or $500\,\Omega$ for the Actel antifuses. This is not completely vindicated in the fan-out related cell delays. Worst-case cell plus routing delays degrade from 3.7 ns to 6.2 ns, for a typical Actel FPGA with fan-out increase from one to eight, while the same figures for QL B-series are 1.7 ns to 4.8 ns.

Much of the interconnect delay is due to track capacitance and Actel, with its smaller logic cell, is likely to use shorter connections between cells so the extra fan-out has a lesser effect than the QL layout, with its large cells. QL's basic cell delay is probably lower because some logic functions are possible with a single level of gating compared with Actel's two-level multiplexer structure.

Stand-by supply current is specified at a maximum of 10 mA for QL FPGAs, relatively high for a true CMOS family. Family details are listed in Table 6.2.

6.3 Xilinx XC8100 FPGAs

6.3.1 XC8100 structure

The Xilinx antifuses, called MicroVias, employ metal/metal connects when programmed. This, like the Quicklogic structure, yields small area connections with low capacitance and low series resistance, leading to an interconnection network with minimal delays. In fact, each extra fan-out adds about 1 ns to the network delay.

XC8100 parts exhibit the finest grain structure of any of the antifuse FPGAs. The basic cells, or CLCs (Configurable Logic Cells), may be configured as a number of basic gates with up to four inputs, or as a D-latch with clear, or tri-state buffer. Unlike the other antifuse families the functions are configured internally; the Actel and QuickLogic logic

Antifuse FPGAs   143

functions are configured by connecting appropriate inputs to ground or Vcc.

The four inputs and one output may be connected to the routing channels but there is also a cascade input and output associated with each CLC. The cascade path is a hard-wireable connection to adjacent cells which by-passes the programmable routing channels and, therefore, suffers a shorter delay than normal connections. It is used to build wide gates, flip-flops and to implement functions like shift registers and fast carry in arithmetic and counter circuits.

Standard routing is accomplished in horizontal and vertical interconnection channels, similar to other FPGAs and ASICs. Each cell has the capability of connecting to 47 horizontal wires. These span either four cells, eight cells or sixteen cells; in addition, there are horizontal long lines which run the whole width of the chip. The long lines are driven from the edge

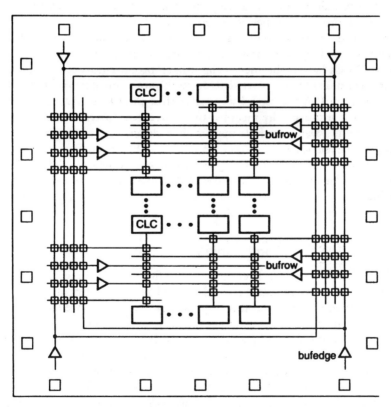

Figure 6.11 XC8100 buffer configuration (reproduced by permission of Xilinx Corporation)

of the chip by row buffers whose fan-out can cope with all the CLCs in the row.

The row buffers are driven, in turn, by one of four dedicated vertical long lines running along each edge of the chip. Their signals are derived from column buffers and are the fastest path for clock signals, when routed directly to the CLC at the end of each row. When used in conjunction with the row buffers it can drive every CLC on the chip, but its source must then be one of four specific external pins. Figure 6.11 shows the buffer configuration for the smallest chip in the family, the XC8100.

The routing resource is completed by vertical channels between the blocks of cells (four cells to a block). Vertical wires exist in equivalent lengths to the eight-CLC and sixteen-CLC horizontal wires, and are used for routing signals between horizontal rows. There are also twelve long vertical wires on each side of the device; these may connect to any horizontal wire.

The I/O cell conforms to the pattern of most other FPGAs, although it does have one or two special features.

The input buffer may be configured, on a global basis, to be CMOS or TTL compatible. There is no registering capability within the I/O cell; internal cells have to be used if input latching or registering is required.

The output buffer has tri-state control, programmable as active-HIGH or active-LOW. A pull-up resistor may also be programmed for each output buffer; it is automatically included on unused pins to avoid floating inputs. There is a novel slew-rate option. For driving transmission lines and capacitive loads, the pull-down transistor can be driven hard to give a sink current of up to 24 mA. As simultaneous switching can lead to ground bounce, a slew-rate limited mode may be chosen. In this situation the pull-down drive is reduced when the output voltage reaches 1 V and ground bounce is significantly reduced.

6.3.2 Design and programming

The XC8100 family has been designed to fit in with logic synthesis design methods, such as VHDL in particular. The basic cell library corresponds in most cases to the primitive functions which are used by VHDL to generate logic functions.

The most efficient way of designing these FPGAs, then, is to use a VHDL compiler or similar synthesis tool. The other standard methods of design entry, such as schematic capture and logic/state equations, are also supported by the XC8100 development system (XDS).

Because the cell and fan-out delays are well defined, they have been built into the synthesis tools. A good estimate of device performance can, therefore, be made before the actual layout phase of the design flow. Actual circuit delays can also be estimated manually from the data sheet values of cell and fan-out timings.

Place-and-route and post-layout timing analysis proceed as with most other FPGA design systems. It is claimed that the abundant routing resources allow for up to 95% utilization of CLCs on any chip. The post-layout timing result should differ very little from the first estimate because the high routability ensures that most inter-cell connections occupy the minimum length.

The end result of the layout procedure is a programming file which instructs the programmer which fuses are to be blown to create the design in silicon. As well as checking that only the correct connections are made by the programmer, it also measures the resistance of each programmed MicroVia. This feature guarantees that device performance will be met as fan-out timing is crucially dependent on antifuse resistance.

6.3.3 XC8100 performance

Xilinx have estimated the performance of XC8100 in several configurations. The raw flip-flop speed, based on minimum clock widths, is 150 MHz, but real circuit configurations degrade this to 123 MHz for a loadable shift register, and as low as 30 MHz for a 16-bit accumulator.

Propagation delay from an input pin to an output pin through a single CLC level is 9 ns.

As yet, no enhanced speed selections have been announced but, with the usual improvements in technology, higher speed versions will probably be available in future.

Zero frequency power consumption is relatively high, ranging from 5 mA for the 384-cell device to 40 mA for the largest family member announced, with 2688 CLCs.

The range summary is listed in Table 6.3.

Table 6.3 *XC8100 family*

Part number	I/O cells	Logic cells	Flip-flops per cell	Inputs per cell	Min. T_{pd} (ns)	Icc (0 MHz) (mA)
XC8101	80	384	0.5	5	2.9	5
XC8103	128	1024	0.5	5	2.9	10
XC8106	168	1728	0.5	5	2.9	20
XC8109	208	2688	0.5	5	2.9	40

7 Selecting and using FPGAs

7.1 Basic criteria

7.1.1 Partitioning

Traditionally, a designer faced with a system specification will take a blank sheet of paper and start to draw boxes representing the functional parts of the system. Typically these will include some intelligence, some memory, some dedicated functions and the I/O interface to the outside world. Some of these boxes will use a single chip, such as a microprocessor, while others may need a whole card of devices.

As technology has progressed many functions have been condensed into single chips so the task of the designer has become increasingly one of selecting dedicated chips and making them talk to each other. The other side of the coin is the task of writing software enabling the microprocessor to control the system.

This process, of breaking a system into its constituent functions, is called **partitioning** and is the first step in creating a **top-down design**. With the advent of programmable LSI and FPGAs, designers have been given far more choice in the way in which functions can be allocated to specific chips. They can include specific functions and the glue logic which is needed to create the interfaces between the dedicated functions.

The designer still needs some basic information before he can allocate functions to an FPGA; he must estimate the logic content, the I/O count and the performance required. Unless he is pushing a particular FPGA family to the limit of performance, gate count or connectivity, the designer will usually have the option of moving the design to a faster selection, larger chip or bigger package if the original estimates fall short of reality.

With so many FPGA families now available, the choice of which to use in any one application can be quite daunting. By analysing the three characteristics for each family, the choice may be narrowed down; the final choice can then be made by reference to the fourth important factor – cost. This is something which must be left for the designer to sort out for himself. Analysing technical factors is an exact science; commercial factors have too many other influences to be subject to precise analysis.

7.1.2 Performance factors

We have already discussed the performance of all the FPGA families covered in the earlier chapters. The difficulty in comparing FPGAs is to ensure that the measurement of performance is specified in the same way for each type. An obvious example is operating frequency. This can be specified as the maximum frequency at which a flip-flop will divide by two, the reciprocal of the sum of the minimum high and low pulse widths, highest frequency at which a counter will operate, or the fastest rate at which a state machine will operate, and so on.

Some divergence also exists with propagation delay. This should be an easy parameter to define, but different architectures throw up different problems. One measurement would be to program a simple gate into the chip and measure, or predict, the delay from input to output, with standard loading on the output. This certainly gives a result which can be derived for any FPGA.

Suppose, though, that we specify a complex gate – a 20-input AND or 16 + 16 AND-OR, for example. A PAL-type device will have the same delay as the simple gate, but any fine-grained FPGA will suffer an increased time penalty because a multi-level solution will be required. An additional level of complexity, requiring multiple passes through a PAL-type array, will penalize this type of device compared with the 'genuine' FPGA.

What we must do, therefore, is specify how the measurements should be made and implement these for every device family. In the next section we will look at a complex suite of benchmarks endorsed by most device manufacturers – the PREP suite – but for the purposes of a simple comparison based on the manufacturer's data let us use four well-defined cases as the basis.

We can justify this because not every family is represented in the PREP benchmark test results and, while the PREP circuits may reflect the structure of real systems we are chiefly interested in comparing perform-ance. By comparing these results with the PREP figures, for the families tested by PREP, we can judge by how much the raw speed figures need to be adjusted to mirror real circuits.

The four circuits we can examine are:

1. Four-input AND gate (input to output delay)
2. Eight-input identity comparator (input to output delay)
3. Shift register (max clock frequency)
4. 16-bit counter (max clock frequency – this is a PREP circuit).

Table 7.1 Comparison of FPGA timings

Family description	4-input NAND (ns)	8-input comparator (ns)	Shift register (MHz)	16-bit counter (MHz)
ACT1 series	17.8	26.8	70	40
ACT2 series	17.0	24.6	125	63
ACT3 series	9.9	14.5	167	82
AT6000 family	9.3	11.7	119	70
ATV family	12	19	100	71
EPX family	10	101	100	83
FLASH370 family	8.5	8.5	143	143
FLEX8000/10K family	14.7	17.1	108	95
pLSI1000 series	10	17.4	113.6	111
pLSI2000 series	7.5	13.6	147	137
pLSI3000 series	17	27.5	71.4	70
MACH 1 series	12	24	76.9	71.4
MACH 2 series	10	20	100	91
MAX5000	15	25	100	83
MAX7000	5	7.4	188	178
MAX9000	16.2	26	117	117
MPA1000 series				
ORCA 2C series	10.1	13.9	119	100
PA7000 series	20	20	71	52
PML series	40	60	20	20
QL B series	5.9	7.6	188	154
XC3100 series	7.2	9.4	95	66
XC4000 series	12.3	16.3	87	45
XC5000 series	19.0	25.0	69	37
XC7300 series	5	8.5	125	125
XC8100 series	11	17.4	123	50

Note 1. Delay through 'built-in' comparator – otherwise 20 ns.

Table 7.1 shows the results of estimating performance, or quoting performance from the various manufacturers data sheets, for a representative device from each FPGA family. The best case for each circumstance is used. For example, in the case of antifuse FPGAs we have assumed a fan-out of one for the input and logic modules. In the RAM-based FPGAs, most estimates of propagation delay have ignored interconnect delays; this may give an optimistic value for delay time but all the RAM-based and antifuse FPGAs have layout-dependent delay times.

7.1.3 Gate and I/O count

Having decided which families are capable of meeting the performance requirement, the choice of device depends on three main factors. These are gate count, I/O count and cost. Cost is a variable which must be found out by the user, and depends to a large extent on market forces. All manufacturers will tell you that they are competitive but, unless you know how to compare their product with the others, you will not be able to judge the truth of any claims.

Gate count can be a contentious issue when dealing with ASICs. Gate arrays are often judged on the basis of two-input gates, as this may be considered as the basic building block of all logic circuits. Thus, a four-input gate has a gate count of three, as has a D-latch, and a D-type flip-flop is, therefore, six gates.

Trying to estimate programmable logic gate is not so straightforward, however. The MACH110, for example, has 32 macrocells, each with four 22-input AND gates (allowing one input per true/complement pair), a four-input OR gate and a D-type flip-flop. A 22-input gate requires 21 two-input gates in cascade, the four-input OR gate another three and the flip-flop six gates – a total of 93 gates per macrocell, or 2976 gates altogether. AMD do not claim that the MACH110 is a 2976-gate device, although they do claim 900 'PLD gates' for the device.

The 'true' gate count must depend on the type of circuit being implemented in the device. For example, if every macrocell was configured as a four-input gate, the MACH110 would contain only 96 equivalent two-input gates. A device such as the A1010B, on the other hand, could be configured with every cell as a four-input gate which would give it a gate count of 885 equivalent two-input gates. Actel claim that this device is a 1000-gate part but, with this criterion, it contains nine times as many gates as the MACH110.

If we look at a more complex circuit, such as a four-bit cascadable up/down counter, the situation changes. A MACH110 needs five macrocells to implement this function, which has a nominal gate count of 66. On this basis, then, the MACH110 contains 6.4 × 66 or 422 gates. The A1010B uses 24 cells per counter so, with 295 cells in total the equivalent gate count is 811, or less than double the MACH.

A function with wide gating shows the MACH in its best light. Consider a circuit with two 16-input decoders OR-ed together; this would use 31 two-input gates and 32 such circuits would fit into a MACH110, suggesting a gate count of 992. The A1010B would need eleven cells to construct this function leading to an estimate of 831 gates.

The PAL-type structure may, therefore, be equivalent to 96, 422 or 992 gates, depending on what type of function it contains; the true FPGA has 885, 811 or 831 for the same cases, a much more consistent estimate. In practice, an LSI device is likely to contain all manner of circuit types so one way to estimate the logic capability of a device is to calculate the equivalent gate count for a number of different functions filling the device, and average these. The functions chosen for this exercise are:

1. Four-input gate
2. Eight-input identity comparator
3. Eight-bit shift register
4. Four-bit up/down binary counter (loadable and cascadable).

These estimates appear in Tables 7.2 to 7.9, where devices are grouped according to package size, and the I/O count and average performance figures are included as well. The FPGA user may find these tables useful as an initial guide to selecting the most appropriate part for the application. The actual gate count and performance will depend on the circuit being designed, and the true fit and performance will only become apparent after device fitter has completed its work.

The tables also indicate which devices are capable of in-circuit programming and support JTAG scan path testing. This last feature will be described later in this chapter.

Table 7.2 FPGAs with 44 pins

Part number	I/O cells	Gate count	Average speed (MHz)	Icc (O MHz) (mA)	JTAG interface	In-circuit programming
ATV2500	38	558	76	180	No	No
CY7C371	32	327	130	175	No	No
CY7C372	32	654	112	250	No	No
EPM5064	28	661	72.5	125	No	No
EPM7032	32	330	175	35	No	No
(is)pLSI1016	36	458	95.5	150	No	Yes
(is)pLSI2032	34	229	122	40	No	Yes
MACH110	32	304	68	150	No	No
MACH210	32	627	85	180	No	No
MACH215	32	304	60	180	No	No
PA7140	38	290	56	150	No	No
XC7236A	34	372	50	126	No	No
XC7318	35	186	142	90	No	No
XC7336	38	372	142	126	No	No

Table 7.3 FPGAs with 68 pins

Part number	I/O cells	Gate count	Average speed (MHz)	Icc (0 MHz) (mA)	JTAG interface	In-circuit programming
A1010B	57	970	51	3	No	No
ATV5000	60	1467	36	350	No	No
EPM5128A	52	1324	72	225	No	No
EPX740	40	413	96	20	Yes	Yes
(is)pLSI1024	54	687	80.5	190	No	Yes
MACH120	48	456	68	180	No	No
MACH220	48	943	85	300	No	No
PML2552	53	349	21	10	No	No
QL8x12B	56	1156	160	10	No	No
XC2064	58	608	70	5	No	Yes
XC7354	49	558	121	140	No	No

Table 7.4 FPGAs with 84 pins

Part number	I/O cells	Gate count	Average speed (MHz)	Icc (0 MHz) (mA)	JTAG interface	In-circuit programming
A1020B	69	1799	51	3	No	No
CY7C373	64	654	112	250	No	No
CY7C374	64	1308	71	300	No	No
EPM5130	64	1324	72	250	No	No
EPM5192A	64	1984	72	225	No	No
EPM7064	64	661	175	50	No	No
EPM7096	72	993	175	80	No	No
(is)pLSI1032	72	916	80.5	190	No	Yes
(is)pLSI2064	68	458	122	tba	No	Yes
MACH130	64	608	60	180	No	No
MACH230	64	1258	60	360	No	No
MACH435	64	1258	81	225	No	No
PML2582	69	349	21	10	No	No
QL12x16B	68	2313	160	10	No	No
XC2018	74	952	70	5	No	Yes
XC3120A	64	1007	101	8	No	Yes
XC4002A	64	1145	68	10	Yes	Yes
XC7272A	60	744	50	252	No	No
XC7372	62	744	121	187	No	No

Table 7.5 FPGAs with 100–133 pins

Part number	I/O cells	Gate count	Average speed (MHz)	Icc (0 MHz) (mA)	JTAG interface	In-circuit programming
A1225A	83	1733	72	2	No	No
A1415A	80	904	105	2	No	No
A1425A	100	1400	105	2	No	No
AT6002	96	3188	95	0.5	No	Yes
AT6005	108	9763	95	0.5	No	Yes
EPF8282	78	1980	107	0.5	Yes	Yes
EPX780	80	827	96	20	Yes	Yes
(is)pLSI1048	106	1374	68	235	No	Yes
(is)pLSI2096	102	687	96	tba	No	Yes
(is)pLSI3192	96	1374	59	tba	Yes	Yes
MACH355	96	943	65	225	Yes	Yes
MACH445	64	1258	65	255	Yes	Yes
MACH446	64	1258	97	255	Yes	Yes
XC3130	80	1574	101	8	No	Yes
XC3142	96	2266	101	8	No	Yes
XC4003	80	1790	68	10	Yes	Yes
XC4004	96	2577	68	10	Yes	Yes
XC5202	88	2434	50	10	Yes	Yes
XC8101	80	1092	80	5	Yes	No

Table 7.6 FPGAs with 134–160 pins

Part number	I/O cells	Gate count	Average speed (MHz)	Icc (0 MHz) (mA)	JTAG interface	In-circuit programming
A1240A	104	3081	72	2	No	No
A1440A	140	2542	105	2	No	No
AT6003	120	4981	95	0.5	No	Yes
CY7C375	128	1308	91	300	No	No
CY7C376	128	1962	75	tba	No	No
CY7C378	128	2616	75	tba	No	No
EPM7128E	96	1324	175	60	No	No
EPM7160E	100	1655	175	90	No	No
EPM7192E	120	1986	175	110	No	No
EPF8452	120	3198	82	0.5	No	Yes
EPF8636	136	4797	82	0.5	Yes	Yes

Table 7.6 *(continued)*

Part number	I/O cells	Gate count	Average speed (MHz)	Icc (0 MHz) (mA)	JTAG interface	In-circuit programming
(is)pLSI3256	128	1832	59	tba	Yes	Yes
MPA1036	120	6596	tba	tba	No	Yes
QL16x24B	104	4627	160	10	No	No
XC73108	108	1117	121	227	No	No
XC3164	120	3525	101	8	No	Yes
XC4005	112	3508	68	10	Yes	Yes
XC4006	128	4582	68	10	Yes	Yes
XC5204	112	3802	50	10	Yes	Yes
XC8103	128	2918	80	10	Yes	No

Table 7.7 *FPGAs with 161–208 pins*

Part number	I/O cells	Gate count	Average speed (MHz)	Icc (0 MHz) (mA)	JTAG interface	In-circuit programming
2C04	160	4193	97	1.5	Yes	Yes
A1280A	140	5545	72	2	No	No
A1460A	168	3820	105	2	No	No
AT6010	173	19925	95	0.5	No	Yes
CY7C377	192	1962	75	tba	No	No
CY7C379	192	2616	75	tba	No	No
EPM7256E	160	2648	175	140	No	No
EPF8820	152	6397	82	0.5	Yes	Yes
EPF10K10	148	7108	82	0.5	Yes	Yes
EPX8160	168	1655	71	1	Yes	Yes
(is)pLSI2128	136	916	96	tba	No	Yes
(is)pLSI3320	160	2291	59	tba	Yes	Yes
MACH465	128	2436	65	tba	Yes	Yes
QL24x32B	180	9254	160	10	No	No
XC3195	176	7618	101	5	No	Yes
XC4003H	160	1790	68	10	Yes	Yes
XC4008	144	5800	68	10	Yes	Yes
XC4010	160	7160	68	10	Yes	Yes
XC5206	148	7463	50	10	Yes	Yes
XC73144	156	1489	94	tba	No	No
XC8106	168	4925	80	20	Yes	No

Table 7.8 FPGAs with 209–299 pins

Part number	I/O cells	Gate count	Average speed (MHz)	Icc (0 MHz) (mA)	JTAG interface	In-circuit programming
2C06	224	6038	97	2.0	Yes	Yes
A14100A	228	6198	105	2	No	No
EPM9320	164	3310	84	90	Yes	Yes
EPM9400	180	4138	84	110	Yes	Yes
EPM9480	196	4966	84	130	Yes	Yes
EPM9560	212	5794	84	160	Yes	Yes
EPF81188	184	9595	82	0.5	No	Yes
EPF81500	208	12337	82	0.5	Yes	Yes
EPF10K20	196	14792	82	0.5	Yes	Yes
EPF10K30	244	20276	82	0.5	Yes	Yes
MPA1064	160	11726	tba	tba	No	Yes
MPA1100	200	18322	tba	tba	No	Yes
XC4005H	192	3508	68	10	Yes	Yes
XC4013	192	10310	68	10	Yes	Yes
XC4025	256	18330	68	10	Yes	Yes
XC5210	192	12337	50	tba	Yes	Yes
XC5215	244	18430	50	tba	Yes	Yes
XC8109	208	7662	80	40	Yes	No

Table 7.9 FPGAs with more than 300 pins

Part number	I/O cells	Gate count	Average speed (MHz)	Icc (0 MHz) (mA)	JTAG interface	In-circuit programming
2C08	224	8218	97	2.7	Yes	Yes
2C10	256	10734	97	3.4	Yes	Yes
2C12	288	13250	97	4.2	Yes	Yes
2C15	320	16773	97	5.0	Yes	Yes
2C26	384	24153	97	7.2	Yes	Yes
EPF10K40	276	28534	82	0.5	Yes	Yes
EPF10K50	308	34943	82	0.5	Yes	Yes
EPF10K70	356	42243	82	0.5	Yes	Yes
EPF10K100	420	61784	82	0.5	Yes	Yes

7.1.4 PREP data

The PREP Corporation (Programmable Electronics Performance Corp.) is an independent body supported by the majority of the programmable logic manufacturers. It has established a set of nine benchmarks which are used to compare the performance and logic density of the different products. The circuit functions are:

1. Data path (multiplexer followed by a direct register and shift register)
2. Timer/counter (loadable 8-bit counter with registered input and output compare)
3. Small state machine (eight states with 13 transitions)
4. Large state machine (sixteen states with 40 transitions)
5. Arithmetic circuit (4 × 4 multiplier driving an accumulator)
6. Accumulator (16 bits)
7. Counter (16 bits with load, count enable and asynchronous clear)
8. Counter (16 bits, prescaled with delayed count after load)
9. Memory map (16 bits with changes at active clock edge).

Details of these circuits may be obtained from PREP Corp. or any of the supporting vendors.

A device under examination is designed with multiple instances of each circuit function. The way in which each function is repeated is specified in the description of that circuit. Figure 7.1 shows how this is arranged for the arithmetic circuit (benchmark 5).

The timing performance is estimated by the worst-case delay of its slowest path. In the case of internal frequency it is taken as the longer of the delays between two flip-flops in the same instance, or two flip-flops in successive instances, including any combinatorial delays. The external frequency also includes delays between flip-flops with external connection between the last and first instance, or the delay between control signals and a flip-flop in the first instance (including set-up time). The worst-case, best-case and mean must be reported for the internal frequency estimate. If the circuit may be optimized for either speed or size, PREP allows for both cases to be reported.

The results for each case give an indication of the performance of the device under consideration in a real circuit. The temptation may be to average the nine figures, or eighteen if both internal and external results are included. The overall average has no real meaning in terms of the performance of a real circuit; however, studies have shown a good correlation between the average and the spread of performance in real designs.

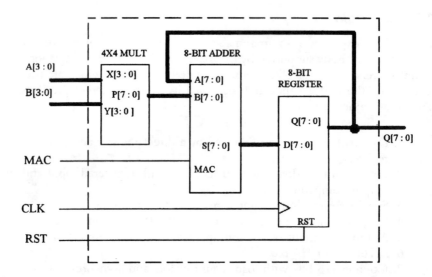

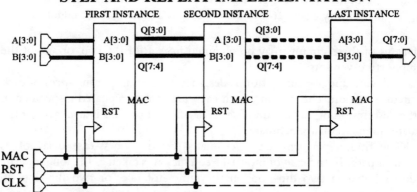

Figure 7.1 *Arithmetic circuit multiple instances showing step-and-repeat implementation (reproduced by permission of PREP Corporation)*

Table 7.10 gives the averages of nine internal and nine external estimates for devices covered by PREP, with the simple figure obtained from the four circuit analysis above as comparison. The correlation, plotted in Figure 7.2 for the internal timings, shows that our estimates are generally higher than the PREP figures. This is because we have included purely combinatorial circuits in our analysis. The table also shows that true FPGA devices show up poorly in the external result; this is because the I/O cell delay has an important influence in this figure, compared with PAL-type parts.

Table 7.10 PREP performance averages

Device number	Internal average (MHz)	External average (MHz)	'Local' average (MHz)
A1240A-2	48.4	23.3	52.3
EPF81188-3	39.8	27.8	82.4
EPM7128-10	73.8	61.6	96.0
EPX780-10	57.2	57.2	83.2
pLSI1048-80	61.2	41.0	67.7
MACH230-15	45.4	37.4	43.1
QL8X12A-3	68.0	44.0	96.8
XC73108-10	52.4	36.0	78.6

We can perform a similar exercise for the logic density estimates. The PREP method simply counts the number of instances of a benchmark circuit which can be fitted into the target. The average, in this case, probably has more meaning, and the total of the nine benchmarks is listed in Table 7.11 together with our estimate of gate count for the largest device in each family. The correlation is much better with most families lying close

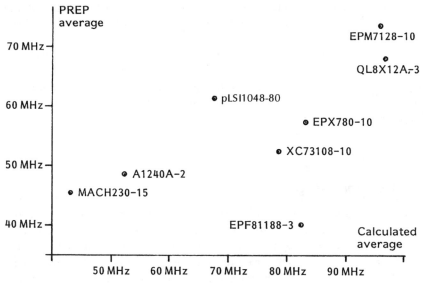

Figure 7.2 Correlation between PREP timings and simple timing analysis

Table 7.11 *PREP capacity comparison*

Device number	PREP Total instances	Estimated gate count	Gate count per instance
A1280A	258	5545	21.5
A1425	61	1400	23
EPF81500	556	12337	22.2
EPM7192	100	1986	19.9
EPX780	36	827	23
pLSI1048	70	1374	19.6
MACH230	43	1258	29.2
QL12X16	73	2313	31.7
XC73108	53	1117	21.1

to an average of 23 gates per benchmark total (actually 207 gates per benchmark). This should give reasonable confidence in our capacity estimates in Tables 7.2 to 7.9.

PREP is clearly a valuable tool for helping to choose the right FPGA for an application and has the advantage that it is supported by the major FPGA manufacturers. Some care should be taken with the interpretation of the results, particularly when this is undertaken by a manufacturer in promotional material. Other factors, such as device cost and ease of use must also be taken into account and, of course, the *caveat* about 'lies, damned lies and statistics' should be heeded.

As a final guide to device selection, every device mentioned in this book has been tabulated in Table 7.12 with I/O count and gate count as parameters in a scatter diagram. Thus, given an estimated requirement of these parameters, the reader can immediately find the devices which will fit those requirements.

7.2 FPGA testing

7.2.1 Vector testing

Vector testing was mentioned in Chapter 3, where we saw that test vectors are usually produced as a result of simulation. A test vector is, simply, a set of input conditions together with the output conditions which they generate. Ideally, every circuit node should be exercised by the test vectors; this brings up the notion of testability and test coverage.

In order for a circuit to be 100% testable, every node must be observable; that is, if a node is faulty it must cause an output to assume an incorrect

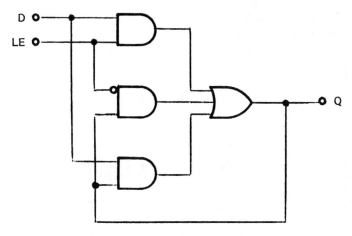

Figure 7.3 D-latch with glitch protection

level with some input combination. The classic example of an untestable circuit is a D-latch with glitch protection; this is shown in Figure 7.3. The bottom AND-gate (D & Q) is only needed to prevent the glitch which can occur when LE changes from HIGH to LOW, with D HIGH, if the gate delays are not equal.

If the connection between the bottom AND-gate and the OR-gate is stuck at LOW, it is impossible to detect this by applying signals to the inputs, unless a glitch is produced. But, in general, tests are only measured when dynamic conditions have stabilized.

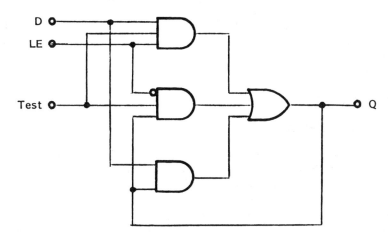

Figure 7.4 Testable glitch-free D-latch

Table 7.12 FPGA selection chart

Gates I/Os	-399	400-599	600-899	900-1349	1350-1999	2000-2999	3000-4499	4500-6999	7000-9999	10000-14999	15000-24999	250000+
300-											2C15	2c26 EPF10K50 EPF10K70 EPF10K100 EPF10K40
250-299										2C10 2C12	XC4025	
210-249								2C06 A14100 EPM9580	2C08		EPF10K30 XC5215	
180-209					CY7C379	CY7C379	EPM9400 XC4005H	EPM9480	EPF81188 QL24×32 XC8109	EPF81500 XC4013 XC5210	MPA1100	
150-179					EPX8160 XC4003H XC73144	EPM7256 pLS13320	2C04 A1460 EPM9320	EPF8820 XC8106	XC3195 XC4010	EPF10K20 MPA1064	AT6010	
125-149				CY7C375 pLSI2128	CY7C376 pLSI3258	A1440 CY7C378 MACH465 XC8103		A1280 EPF8636 XC4006A XC4008	XC5208 EPF10K10			
100-124			pLSI2096	XC73108	A1425 EPM7160 EPM7192 pLSI1048		A1240 EPF8452 XC3064 XC4005 XC5204	AT6003 MPA1036 QL16×24	AT6005			

Table 7.12 *(continued)*

Gates	-399	400-599	600-899	900-1349	1350-1999	2000-2999	3000-4499	4500-6999	7000-9999	10000-14999	15000-24999	250000+
80-90			EPX780 EPX880 XC7372	A1415 EPM7128 MACH 335 XC8101	A1225 pLSI3192 XC3130 XC4003	XC3042 XC4004 XC5202	AT6002					
60-79	PML2852	pLSI2064 XC7354	CY7C373 EPM7064 MACH130 XC7272A	CY7C374 EPM5130 EPM7096 pLSI1032	A1020 ATV5000 AEPF8282 EPM5192A MACH230 MACH435 XC2018 XC3020 XC4002A	QL12×16						
40-59	PML2552	EXX740 MACH120	pLSI1024 XC2084	A1010 EPM5128 MACH220 QL8×12								
-39	CY7C371 EPM7032 pLSI2032 MACH110 PA7140 XC7236A XC7318 XC7336	ATV2500 pLSI1016	CY7C372 EPM5084 MACH210 MACH215									

To make this circuit testable we can add a TEST signal to the upper two AND-gates, as in Figure 7.4. If the latch is set HIGH and then TEST taken LOW, the output will only stay HIGH if the bottom gate is functioning correctly.

Testability is tedious to quantify in a complex circuit. The procedure is first to generate a complete set of test vectors, which begs the question of how they can be defined as complete. Next each node in the circuit is held at either a HIGH level (stuck-at-one or sa1) or a LOW level (stuck-at-zero or sa0), and the output examined to see if there is any change from a perfect device. If there is a change then the node is observable and has been detected.

If any node is undetected, that is sa1 and sa0 produce no change in the outputs, then a test should be devised to detect that node. If that proves to be impossible the node must be considered undetectable. This may be because it is unobservable, as described above, or uncontrollable.

Uncontrollable nodes are those which cannot be changed by making any change to the input conditions. In either case, to provide full testability some circuit changes may be necessary. Once the circuit is 100% testable, the stuck-at fault analysis will show whether the test coverage is also 100%. In an ideal world all circuits should be 100% testable and 100% covered.

For any but the simplest circuits, the stuck-at fault procedure is extremely tedious and time-consuming, but is now part of some CAD programs. Even so, with complex devices the number of tests needed to approach 100% test coverage becomes prohibitive and, in practice, some compromises have to be reached.

While testing is a preferred option for every programmed device it is less important for the reconfigurable parts than one-time programmables. Reconfigurable devices include RAM-based FPGAs and other parts using electrically erasable cells for programming. These devices usually undergo extensive functional testing by the manufacturer to ensure that every macrocell and internal connection is operating correctly. Thus, provided that a device is correctly programmed, the user can have a high confidence that it will function correctly also.

Those devices which use UV-erasable cells in a non-window package, or anti-fuses, cannot be reprogrammed. It seems at first glance that these parts cannot be guaranteed to be functionally good by the manufacturer, but there are ways of ensuring that at least a high proportion of the internal cells and connections are operating correctly.

For example, we have seen that Actel FPGAs use a number of pass transistors to access some of the programming paths. These same transistors can provide a path for testing the logic cells and connecting paths. Another technique is to insert extra fuses which, when intact, allow the internal functions to be tested but, when blown, return the device to its normal unprogrammed state.

The proportion of functional failures of correctly programmed FPGAs is, in practice, very small thanks to these manufacturing techniques. FPGAs must be tested at some stage and, for those which need programming before they are inserted into printed boards, the ideal situation is to test them in the programming socket. Most commercial programmers do allow this, provided that test vectors are specified and appended to the programming file.

FPGAs which are programmed in-circuit must be tested as part of the final system although, as we discuss in the next section, procedures have been developed for achieving this and isolating the circuit elements so that faulty devices can be pin-pointed.

7.2.2 Boundary scan testing

Even if an FPGA is operating correctly by itself it must interface correctly with the other components in the final system and, to this end, attention must be paid to how the total system may be tested. Past practice has been to use 'bed of nails' testing, involving a set of probes from a dedicated test fixture making contact with the tracks on a printed circuit board after assembly. There are a number of drawbacks to this approach.

Firstly the size of components, and their PCB contact areas, is becoming smaller making it more difficult to make contact reliably with the tracks. Secondly, components may be mounted on either side of the board making it necessary to probe from both sides. Thirdly, damage may be done by the probes themselves during the test procedure. Accordingly, the Joint Test Action Group (JTAG) devised a standard for incorporating test structures into the devices themselves. This applies to all components, not just FPGAs.

The principle is to surround the core of the device with a shift register which can be loaded with test data and then read out to confirm that the board is operating correctly. This may be done as a one-off production test or, for example, every time the system is powered-up during normal use. The principle is shown in Figure 7.5.

The full implementation of JTAG boundary scan to IEEE1149.1–1990 involves the addition of several components to a device, also shown in Figure 7.5.

The interface is the test access port (TAP) which contains the four I/Os needed to control boundary scan operation and, in particular, a state machine controlled by the test mode select (TMS) signal. This sends control signals to the registers which form the body of the scan circuit. These registers are the Instruction Register, the By-pass Register, the Boundary Scan Register and the Identification Register.

The instruction register holds data which determines the function which is being carried out. For example, BY-PASS allows data to pass through the

164 *FPGAs and Programmable LSI: A Designer's Handbook*

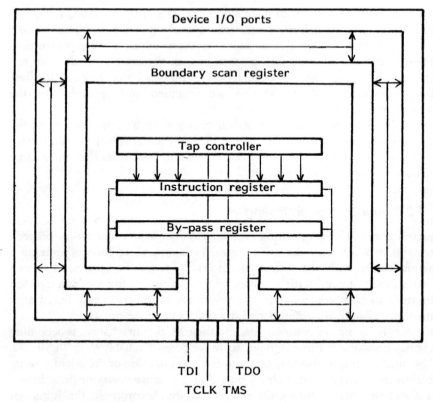

Figure 7.5 *JTAG test structure*

circuit without affecting normal device operation. EXTEST forces data onto the device outputs and receive data on the device inputs, for the purpose of checking connections between chips. Other modes include SAMPLE/PRELOAD for loading the boundary scan register during normal device operation, and RUNBIST which allows a user register to be loaded with test data for testing the device core logic.

A full description of the JTAG functions may be found in the data sheets of the appropriate devices, describing how each fulfils the specification. What is not necessarily specified in detail is the overhead involved in meeting the JTAG requirements.

There are three ways in which JTAG may be implemented. It may be built into the chip as a permanent feature, and additional to the programmable core of the chip. In some cases it must be designed as part of the circuit, in which case it reduces the amount of user logic which can be included with the boundary scan. The other way in which it can be included is by designing it as a RAM-based FPGA. JTAG may then be programmed as a test mode circuit and the chip reprogrammed with the desired circuit once testing is complete.

In the first case the chip area, and hence cost, will be higher than a non-JTAG circuit but this is inevitable if boundary scan is essential to the system. If the JTAG circuit has to be built into the FPGA, the logic used is no longer available for user circuitry. This overhead has been estimated as being as high as 33% in an A1280; this is of the order of 1800 gates. It will also affect the performance of the FPGA as there are multiplexers on the I/Os to route test data into the JTAG area.

The RAM-based solution has obvious advantages over the other two methods. The only overhead is design effort, which can be re-used on subsequent designs with the same device, and the time taken to test the system and reload the user circuit, if a full system test is performed. This method does have the drawback that the chip is not tested with its final program in place. If that function is required the JTAG interface will have to be included as an extra function and the overhead borne as a necessary added cost.

If a system is being designed for JTAG testing, the choice of FPGA will be influenced by this requirement. Also it should be noted that the only mandatory modes are BYPASS, SAMPLE/PRELOAD and EXTEST, thus, if internal testing is required, for example, it is necessary to check that the chosen device supports the appropriate modes.

Boundary scan JTAG testing, then, can be considered on two levels. As a basic feature it allows the overall system to be checked for proper connectivity and assembly. That is, the right parts are in the right places and all the pins are connected together, as specified. In the advanced option, each device can be tested and any faults pin-pointed, giving a high level of confidence that the whole system will function correctly when it, too, is tested.

7.3 Programming FPGAs

7.3.1 Machine programming

Traditionally, PLDs have needed to be programmed before insertion into the printed circuit board. This is because they have needed voltages in excess of the normal 5 V which powers the standard integrated circuit families. Any attempt to program PLDs in the circuit could damage other devices connected to the same power rail so they must be programmed first. Although some FPGAs have been designed to generate the higher voltages on-chip, the majority of non-volatile devices still need to be programmed out of circuit.

There are two issues concerning programming which need resolving before embarking on production with FPGAs. These are programming yield and programming throughput.

Out-of-circuit programmed FPGAs may be considered in two classes: reprogrammable and one-time programmable. Reprogrammable devices may be tested by the manufacturer and then erased before shipping. This allows both functional testing, as discussed above, and testing of the programming circuitry. Experience shows that the programming yield on reprogrammable FPGAs is very close to 100%; any failures are more often due to machine or socket problems.

One-time programmable FPGAs, which include the antifuse and plastic packaged EPROM processes, cannot be fully tested for programmability before use, for obvious reasons; once a fuse is blown it cannot be remade. Test fuses are usually incorporated into the chip so some limited programming test is possible. This should weed out any catastrophic failures but programming rejects often exceed 1% in production programming. The manufacturers will usually replace failures; Actel, for example, advise that a failure rate of up to 5% is possible and offer a 'no questions asked' credit or replacement facility for this level of rejects.

As we have already hinted, device problems are not the only reason for poor programming yields. The programming voltage for any device is usually specified with a tight tolerance, typically ±5%, often with rise and fall time limits as well. Fast edges can lead to capacitive coupling inside the chip, causing fuses or cells on adjacent lines to be programmed in error. Thus, it is essential that device programmers are kept calibrated to maintain reliable operation.

It is also important that the programming socket is kept clean to ensure good contact with the device pins. The fuses are selected by a row and column addressing scheme so, if a HIGH level is transmitted instead of a LOW because of poor contact the wrong fuse will be programmed. Some programmers do a contact check by measuring the reverse diode that is present on most device pins. If this cannot be detected it assumes that there is a contact problem and aborts the programming. If these 'continuity failures' persist it is worth cleaning or changing the socket, although regular servicing will prevent this situation arising.

The other important precaution is anti-static handling. This subject is well known in integrated circuit technology, but it is worth remembering that FPGAs are as susceptible to static discharge as any other IC Although device pads usually have protection circuitry, this will only afford protection up to a limit of, say, 2–3000 V and it is easy to generate voltages in excess of this by handling insulated objects. The problem is not just the immediate failure of a device but the damage which does not show itself until the device has been operating for some time.

The main impact of programming yield is in the planning of a production schedule. Programmable devices do offer flexibility; if several programs are used on one FPGA, stocking is simplified as dedication can be left to the last minute, and only one part need be ordered to fill many board positions. If yields are not predictable, though, spare devices need to be stocked and rejects returned for credit – both of which add to the cost of ownership of the project. A common solution is to order pre-programmed parts, at extra cost, which removes the uncertainty and the flexibility.

Programming time is the other variable in this equation. Even with automatic handling equipment, throughput can vary from a few hundred devices per hour down to thirty or less. Unless a throughput of several thousand per week is planned it would not be economic to use automatic handling, so manual programming would have to be used. Even as few as 200 FPGAs might, therefore, add an extra day to the production time and as much as 50p per device to the cost just for labour charges and overheads.

Again, buying pre-programmed devices or using a sub-contract programming service are common solutions – but at a price.

7.3.2 In-circuit programming

RAM-based FPGAs are the usual type of device which may be programmed in-circuit, because of the high programming voltage required by the other technologies. However, some non-volatile devices can be programmed with an external 5 V source. This may be by using an actual 5 V programmable cell, as in FLASH ROMs, or by generating higher voltages on-chip from the 5 V rail.

Programming in-circuit overcomes the problems described above, while maintaining flexibility, but does incur some overhead. Yield is not usually an issue because the programming cells can be pre-tested, but there is still a time penalty and, possibly, a board area penalty. Any programming or functional failures will cause a much higher extra cost because a board rework will be involved, unless the FPGA is socketed.

The time penalty for non-volatile FPGAs is the same as for machine-programmed devices, and some kind of interface has to built onto the board to allow the programming data to be fed to the device. The data is usually on a serial interface, so the number of connections is not large, but they will add to the complexity of the tracking as the programming pins may act as I/Os in non-programming mode.

RAM-based FPGAs have to be programmed every time the system powers up so, although there is no overhead in production, it adds to the system 'warm-up' time. This may not be a problem, depending on the environment. There must be a significant board overhead though, as the program has to be stored in some non-volatile form before it can be loaded into the FPGA. If the system can boot without a functioning FPGA, and has a disk drive, the data can be downloaded from disk. This is an ideal situation; more often an additional EPROM must be used to store FPGA configuration data.

One advantage, for larger package FPGAs, is the reduced handling with in-circuit programming. This reduces the chance of both physical damage and static damage and, given that the other disadvantages can be minimized, offers a significant benefit in production by allowing the FPGA to be treated as any other device.

7.3.3 Device security

Having spent several weeks or months designing a system, the last thing anybody wants is for it to be copied and marketed in competition. While this may be illegal, actually preventing it or obtaining redress can be an expensive and lengthy exercise. Many programmable devices, including PLDs, FPGAs and microcontrollers, but not EPROMs, have a security cell or fuse which prevents data from being read out after programming.

Part of the programming process is a verification; this compares the programmed data with the source data to verify correct programming. Once this process has been completed the security bit can be set and, provided that the program in the device is identified, the device need not be read again. This is not possible with RAM-based FPGAs if the data is stored in an EPROM as the EPROM can be copied and the FPGA loaded from the copy EPROM.

There are ways round this. A secure microcontroller could be used in place of the EPROM. A small program would provide the intelligence to interface to the FPGA and the rest of the memory could store the configuration data. Alternatively a small, secure, PLD could be used to unscramble data from an EPROM stored in scrambled form.

Being able to reproduce the configuration data does not mean that it can be understood. Thus, although it may be possible to copy a RAM-based FPGA design exactly, it would be practically impossible to make even the smallest change to the design because of the impossibility of back-tracking from configuration data to the design parameters.

While reading back a secured device is impossible via a device programmer, more sophisticated methods may be used in an attempt to discover the program. These will be mentioned, not as an aid to making illegal copies, but to help proprietors to safeguard their designs against these methods.

Small devices, with not many inputs, can be examined with a logic analyser while different input combinations are applied. It may then be possible to decipher the function within the device. This is only effective in combinatorial circuits without feedback and even a 16-input device would need 65 536 logic combinations for a full decode. Feedback would complicate this and using registered parts in a state machine make the process unmanageable.

If a device is decapsulated, so that the bare chip is visible with connections intact, scanning electron microscopic analysis can reveal which cells are programmed in electrically and UV-erasable devices. If the geometry of the device is known, or can be deciphered, it is then possible to recreate the original program.

Antifuses are the most difficult to detect. These are small enough to be entirely covered by the metal tracks and there is not usually any visible sign of programming. In some cases a thermal analysis will detect which fuses are passing current and, then, it may be possible to recover the design.

It is practically impossible to prevent anyone determined enough from copying an FPGA. By making it as difficult as possible it may not prove worth the effort to the would-be counterfeiter, but the last resort must be a legal one.

7.4 Migration to ASIC

The main advantages of FPGAs over masked AISCs are reduced time to market and increased flexibility.

It takes a matter of hours, or less, to move from a successful logic simulation to a working FPGA sample, but a week or more to obtain a working masked ASIC. The cost of FPGA iterations is also negligible compared with the cost of masked ASIC iterations.

Planning throughput is much easier with FPGAs, as well. A sudden ramp in production, or a hold, can be managed with an off-the-shelf component, but would cause difficulties with products having a long manufacturing cycle from the point of ordering.

FPGAs are, therefore, the ideal vehicle for the early production phases of a project. If quantities are never going to be high, they may well last right through the life of the project. The chief drawback of FPGAs is cost. Masked ASICs are, usually, much cheaper for a comparable size and performance. At some stage, then, it may prove cost-effective to make the change from an FPGA to an ASIC.

Some FPGA manufacturers offer mask-programmed versions of their programmable parts, and this probably offers the best chance of a seamless transfer. Many ASIC manufacturers will produce equivalent devices from the FPGA netlist and simulation files. They will generally offer a refund of the non-recurring charges if their device does not perform as the original.

The main performance difference is likely to be in timing parameters. Masked ASICs are usually quicker than FPGAs because their connections are metal/metal vias rather than some programmed cell, and will have a lower resistance and parasitic capacitance.

The most difficult situation is replacing an in-circuit programmed FPGA. In this case, some minor board modification may be required to remove the configuration data source. If, though, the FPGA circuit configuration was changeable during system operation, for testing or a Cache Logic application, the transfer to ASIC may not be possible at all.

If the change is seamless, it is not necessary to change 100% of production to an ASIC. If the production volume is uncertain a proportion only need be sourced as ASIC, depending on the minimum requirement. Any excess can then be mopped up with FPGAs. The average cost will still be lower than 100% FPGA, although not as low as if all the production were ASICs.

For large volume projects an FPGA can be used as a development tool with many of the advantages of ASICs, but without the high initial costs and leadtimes. If an appropriate FPGA is chosen, a change to ASIC can be made with little upheaval to the production schedule, and the FPGA option retained for smoothing out any wrinkles.

8 Using FPGAs

8.1 Design techniques

8.1.1 General methods

We covered the basic design methods for FPGAs in Chapter 3; these were logic/state equations, logic synthesis (VHDL) and schematic capture. In general, the methods used for FPGAs are similar to those used for discrete logic, except that FPGA macros tend to be simpler than many of the functions available in discrete logic families.

This lower level of design means that there can be more work involved in designing an FPGA but the process is more flexible. For example, discrete logic families may have devices with four identical gates (quad two-input NAND) but an FPGA designer can select a single gate at a particular circuit location. Making full use of a quad gate can cause difficulties in layout, where the four elements may be used in widespread locations, but this type of problem does not arise in FPGA designs.

Close timing matches can also be a problem in discrete logic. Although all discrete logic circuits are well specified in terms of timing parameters, usually this involves only the maximum propagation delay, or minimum clock width; the best case and spreads are not usually published. The actual timing depends on process parameters such as sheet resistance and oxide thickness, which can vary from batch to batch. Two devices from different diffusion batches may have quite different timing characteristics.

In production, it is quite likely that devices which would ideally be matched might even come from different manufacturers. In FPGAs, all the cells will have approximately the same timing performance so matching is inherently better than with discrete logic. Most design systems allow manual placing of cells so components needing close matching can be placed next to each other on the chip. The most likely source of mismatch is a routing track length difference. While this is less easy to manage, the post-layout timing analysis will point out any discrepancies; although difficult to correct, they will, at least, be predictable.

8.1.2 Fan-out limitations

Discrete logic designers are used to fan-out restrictions. The old 74-series TTL families had a fan-out limit of ten because of the DC loading of inputs and drive capability of the outputs. The VOL was specified at 16 mA and each input needed 1.6 mA of input current when taken LOW. Of course, typically an input would source less than 1.6 mA and VOL would stay below 0.4 V even when sinking more than 16 mA. Circuits breaking the rules would work 99.9% of the time but, on a hot day with a high supply voltage and an unfortunate coincidence of low resistance loads, they would fail. To be safe, designers had to stick to the rules.

With the advent of CMOS and pnp inputs, DC fan-out restrictions became less critical but AC fan-out related performance remained relevant. Timing parameters are specified into a standard load and, if the capacitance of the actual load exceeds the specification, they will be degraded. The capacitative load on an output increases with the number of inputs it is driving and the length of PCB track connected to it.

Cell outputs inside FPGAs suffer from the same restrictions. Some manufacturers limit the fan-out of individual cells; Actel's Designer system puts out a warning for fan-outs greater than ten, and will not allow more than twenty four. All cellular FPGAs show some degradation of delay times as the fan-out and connected track increase. Most have one or more 'clock' lines for distributing high fan-out signals but many circuit designs will include several signals driving more than ten inputs; note that soft macros often have inputs with fan-in greater than one, so it is possible to exceed fan-out guide-lines even when only a few macros are connected to a signal.

The obvious way to overcome this problem is to make a 'buffer tree', as in Figure 8.1, which multiplies the signal and increases the fan-out by a factor equal to the number of buffers. Thus, a fan-out limit of ten can be increased to one hundred by a single level of buffering, or a thousand by adding a second level. The only drawbacks are that cell usage is increased, and the extra level of buffering adds to the delay.

Not much can be done about the first drawback. The second is not so bad as the following example shows. The cell delay for a fan-out of three for ACT1 FPGAs is 3.3 ns, while it is 10.2 ns for a fan-out of eight. Two levels of fan-out three thus has a lower delay than one level of fan-out eight and, provided that utilization is not critical, would yield a higher performance solution.

A better solution is to use parallel logic, particularly if the function is being generated by a single cell. As its name implies, the function is

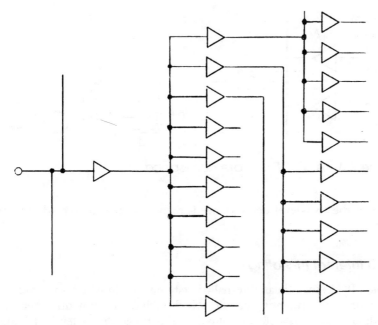

Figure 8.1 *Buffer tree*

replicated as many times as necessary to provide the required level of drive. Even if several cells are used to generate a function it may be possible to produce multiple versions of the function without excessive use of additional cells. Figure 8.2 shows how a 16-bit address may be decoded to drive the sixteen enables of a bus output with a fan-out of four, using three extra cells with only four inputs available for each cell. Only the second

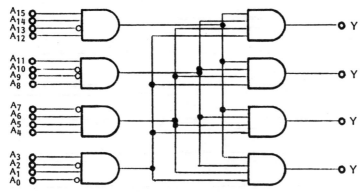

Figure 8.2 *Sixteen-bit address decode with high fan-out*

174 FPGAs and Programmable LSI: A Designer's Handbook

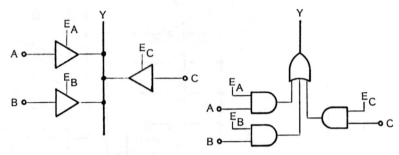

Figure 8.3 AND-OR/tri-state correspondence

stage of the two-level decoder needs to be repeated to provide the extra drive.

8.1.3 Internal tri-stating

Some FPGAs include an internal tri-state option, in which case there is no problem in implementing circuits which include this function. Apart from enabling busses, tri-state can also be used for an extra level of logic by means of the wired-OR function.

In fact, tri-state is only an alternative implementation of the AND-OR function. It is useful in distributed systems because the inputs to the OR part of the function can be separated physically along a bus. Figure 8.3

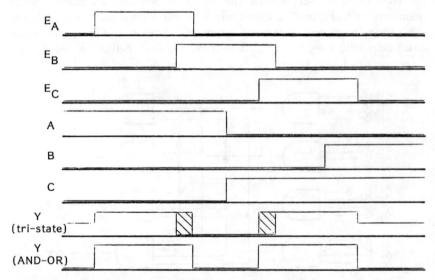

Figure 8.4 Timing diagram for AND-OR equivalence of tri-state

shows the correspondence of AND-OR and tri-state. The tri-state output enables perform the same function as the control input to the AND-gates; the bus acts as a distributed OR-gate.

This logic may be exactly reversed to implement an internal tri-state function in an FPGA which does not have this built-in function. Signals connected to the input of the tri-state buffer are, instead, connected to one input of the two-input AND gate and the output enable to the other input. The outputs from each AND gate are then OR-ed together. This also overcomes any timing problems when two enables overlap and some of the bus lines are momentarily connected to both a HIGH and a LOW. This could damage outputs in a true tri-state environment but will merely produce some uneven timing in an AND-OR circuit, as Figure 8.4 shows.

Often, the enables may be encoded into binary form; in this case a multiplexer can replace the discrete gate solution.

8.1.4 Clock enabling

Because only a few high fan-out lines are available in most FPGAs, for clock distribution, the designer may decide to generate local clocks from the global clock. This has the advantage that all flip-flop outputs are synchronized which can reduce problems with metastability and other asynchronous effects.

A common way of generating a local clock is to gate the global clock with a local enable signal, as in Figure 8.5. If the enable signal changes (LOW to HIGH) while the clock is HIGH, a spurious clock pulse will be generated. This could have a disastrous effect in counters and similar circuits.

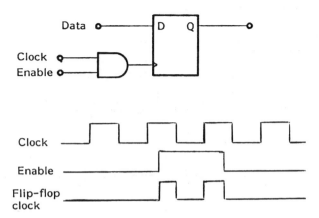

Figure 8.5 *Clock enabling and timing*

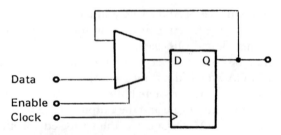

Figure 8.6 Enable flip-flop

A better way to enable flip-flops is to use a dedicated 'enable flip-flop'. Many cell libraries include this function; it is generated from a flip-flop and a multiplexer, wired together as in Figure 8.6. The count enable line is the multiplexer select input; when the flip-flop is enabled the data input is fed to the flip-flop, otherwise the output is fed back so that there is no change when the flip-flop is clocked.

This also eliminates the delay which the clock would suffer if gated directly with an enable signal.

This configuration is particularly useful in counter circuits. Any stage of a counter must 'hold' unless the previous stages are all HIGH (up count), so this signal can be fed directly to the enable input. The inverted output is taken to the data input to provide the inversion or, it too may be multiplexed with a direct input in the case of a loadable counter. A counter stage using an enable flip-flop is shown in Figure 8.7.

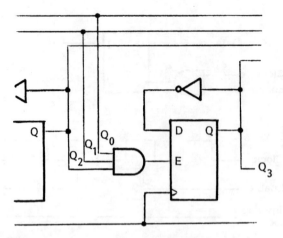

Figure 8.7 Counter stage with enable flip-flop

8.1.5 Optimization

In some circuits the designer may have the option of optimizing for speed or for utilization.

If speed is the ultimate goal of the design, the aim must be to minimize the number of logic levels for critical signals, other things like fan-out being equal. We can illustrate this by using an up-counter as an example, although the principle applies to many other circuits.

The limiting path is the delay between the previous stage going HIGH and enabling the flip-flop. This splits into sections as clock output delay (previous stages), logic delay and set-up time. In an up-counter the count sequence for toggling the nth stage is 0111...01, 0111...10, 0111...11, 1000...00. Thus, it is the transition of the first stage from 0 to 1 which triggers the toggling of all subsequent stages.

Many FPGAs are limited to four inputs per gate, so decoding more than four outputs for an enable signal will require two levels of gating. Because the 1's from the higher stages are established at least one clock cycle before the first stage, two levels are not a problem for these; the important signal is Q0 which must enter the chain at the last level, as in Figure 8.8.

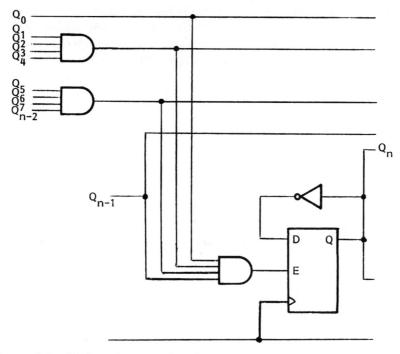

Figure 8.8 *High order counter stage*

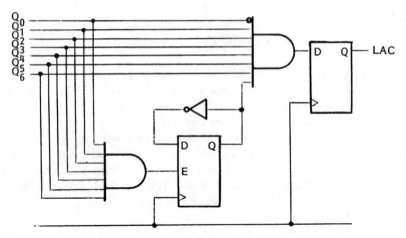

Figure 8.9. Counter look-ahead-carry

In a counter with many stages, the fan-out of Q0 becomes so high that this becomes the limiting factor. Parallel buffering is not an easy option because that would mean duplicating the first counter stage. One possibility is to anticipate the 'all HIGHs' condition by decoding 11111110 (for eight stages) and buffering this for one clock cycle with a single flip-flop. This 'look ahead carry' is then available at the same time as Q0 and allows the next eight stages to operate with the same timing as the bottom eight stages. This is illustrated in Figure 8.9.

This technique may be continued for even higher stages than sixteen and is also applicable to arithmetic circuits. In general terms, it means

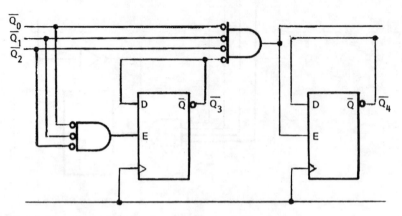

Figure 8.10 Carry chain for minimum utilization

determining which is the most critical signal(s) and gating this at the last stage before the next logic block.

If the counter speed is not critical the design may be optimized for utilization, to use the least possible number of cells. The actual design will depend on the complexity of the logic cell. In general, the technique for the carry chain would be to parallel as many previous stages as possible, then carry this forward as a carry-propagate to the next n stages. This carry chain is shown for the fourth and fifth stages of a counter using a maximum of four inputs per gate, in Figure 8.10.

This will result in, for example, six propagation delays in a 24-stage counter optimized for cell usage compared with a single level of delay in a design optimized for speed. Clearly, the designer has more degrees of freedom in designing FPGAs than when using discrete logic.

8.1.6 State machine implementation

In Chapter 3 we designed a simple state machine, whose function was to decode four digits entered serially, unlock a door if they are correct or sound an alarm if they are not. The physical structure of the machine was a four-bit register driven by some combinatorial logic, which determined the next state, and driving a decoder to derive the present state from the state register.

This structure is derived from smaller PLDs, FPLSs in particular. Most PLDs have no more than sixteen flip-flops in their state register, many have only eight, so some kind of binary encoding scheme is necessary to be able to define more than sixteen states. Our example has ten states, which is why we used a four-bit register for encoding.

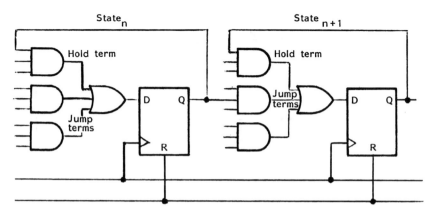

Figure 8.11 *OHE state machine architecture*

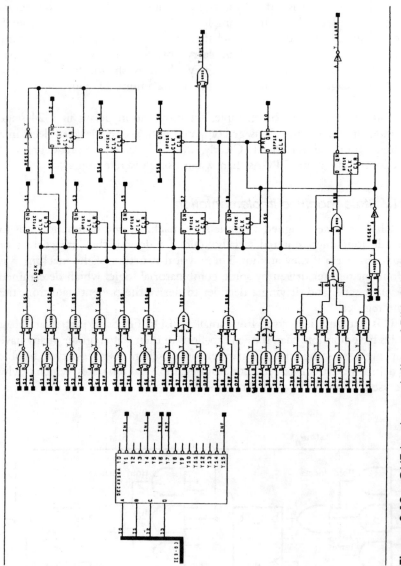

Figure 8.12 OHE implementation of PIN detector

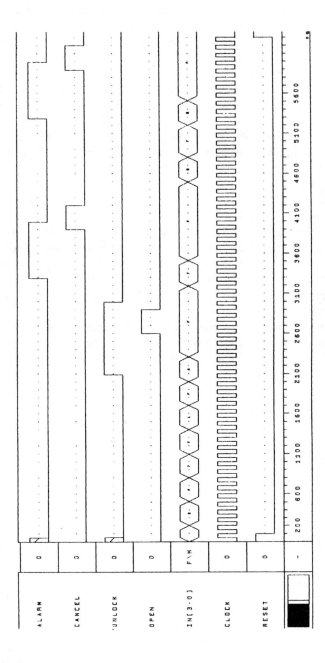

Figure 8.13 *Back-annotated simulation of OHE PIN detector at 10 MHz*

FPGAs generally have a hundred or more flip-flops available so it is often more efficient to allocate one flip-flop per state, connecting the flip-flops in a shift register arrangement. The present state is defined by the appropriate flip-flop being set HIGH, all the others remaining LOW. Because just one flip-flop is active this structure is called one-hot encoding (OHE). Figure 8.11 shows the generalized OHE structure.

The OHE version of the doorlock circuit is shown in Figure 8.12. By eliminating the next state encoder and present state decoder, but adding six flip-flops and two buffers (to maintain fan-out below ten), the module count is reduced from 68 to 64. This is not an enormous saving; indeed, larger state machines might even need more resources in OHE format, but there is a significant performance improvement.

In the binary encoded version there were between six and nine logic levels in the loop; the OHE version has from three to five levels. The state machine runs quite happily at 10 MHz, as the back-annotated simulation in Figure 8.13 shows, and the clock to output delay is reduced to 40 ns.

One danger with one hot encoding is that the HIGH might get 'lost'. It is, therefore, essential that all the HOLD conditions are implemented and that there is a way of returning to a known state. Typically, a NOR gate with an input from every state and the output to the known state will cater for this.

Alternatively, more than one state may become active if two jump conditions are true simultaneously because of a design or timing problem. This situation is more difficult to correct in hardware. An OR gate for each state with that input complemented and the other states true would detect a fault, but this would impose an unacceptable hardware overhead for even modest state machines. The solution is in careful design and good simulation.

8.1.7 Metastability

Metastability may occur when data arrives at the D input of a master-slave flip-flop in violation of the setup and hold times. In particular, if the data goes HIGH just before the clock, a short HIGH pulse will appear on the master section output when the propagation delay through the master is greater than the time gap between data and clock. This pulse may then propagate itself round the master latch and into the slave latch as a high frequency oscillation. Capacitance on the flip-flop output may smooth this into an intermediate level so that it remains in an undefined state for some time.

The phenomenon is usually investigated on a statistical basis, because the outcome is not entirely predictable. There is a metastability window, within the setup and hold period, when the output is likely to suffer, but the final state of the output may be HIGH or LOW. A circuit for detecting

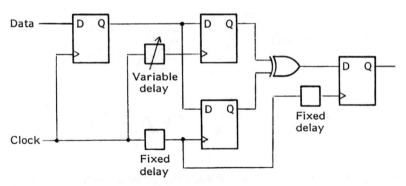

Figure 8.14 *Metastability detection circuit*

metastability is shown in Figure 8.14; the flip-flop under test receives data at random intervals relative to the clock edge – under normal conditions both sampling flip-flops should have the same output – if they are different, as detected by the exclusive-OR gate, a metastable event has occurred.

If the shorter delay is set to a defined settling time (t_{MET}), and the longer made well in excess of this, the MTBF (mean time between failure) can be established by counting events for a known period. When this has been carried out for a number of t_{MET}s, a graph of ln(MTBF) against t_{MET} will characterize the flip-flop and enable failure rate predictions to be made for any settling time. Alternatively, it can give an indication of what allowance must be made for metastable settling to achieve a desired reliability. Typical figures for FPGA processes are 3–4 ns for one-day MTBF and 4–6 ns for one year.

In normal synchronous circuits, metastability should not be an issue as timings can be calculated to ensure that data never changes within a setup time of the clock. The usual area for problems is in synchronization circuits, when an asynchronous signal is clocked into a register. In this case there is no way of predicting when the data might change. A useful technique to improve matters is double synchronization, using opposite clock edges. One flip-flop, or latch, with negative edge trigger is followed by a second one using positive edge trigger. The overall probability is then the product of the two probabilities for the individual flip-flops, which will be vanishingly small.

8.2 Board-level considerations

8.2.1 Ground bounce

In general, the techniques for laying out printed circuit boards for systems containing FPGAs are the same as for any high speed integrated circuit.

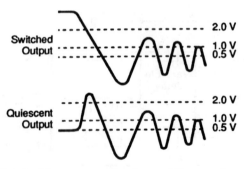

Figure 8.15 *Ground bounce effect on quiescent outputs*

There are some problems which are particularly noticeable in FPGAs; among these is ground bounce.

Ground bounce occurs when several outputs switch simultaneously from HIGH to LOW. The ground connection of any IC has an inductive component and the output has a capacitative element to it. The discharge current from the output capacitance has to pass through the ground connection, and this gives rise to the ground bounce phenomenon.

The discharge current, I = Co.dV/dt, where Co is the output capacitance and dV/dt the slew rate. The current surge produces a voltage surge across the ground lead inductance of L.dI/dt. Because this is a local effect; quiescent outputs in the same chip will also be lifted by this amount but the ground voltage will remain unchanged at devices elsewhere on the board. This may result in spurious switching on quiescent LOW outputs. The effect is illustrated in Figure 8.15.

The size of the ground bounce voltage depends on three factors – the output capacitance, the slew rate and the ground lead inductance – and can be quantified by combining the two expressions above into $L.Co.d^2V/dt^2$. This is a classical decaying sinusoidal form when driven by a step function.

Output capacitance and ground lead inductance are both factors determined by the printed board layout and are minimized by keeping lines short and using a ground plane. Many FPGAs have multiple ground pins and, if possible, the effect can be minimized by using pins near ground pins as outputs, and spreading them so that the simultaneous switching pins do not all share the same ground pin. Using a socket, rather than soldering directly to the board, will also increase these parameters.

Many FPGAs also have slew rate limit option on the outputs. A high slew rate has obvious attractions when driving capacitative loads but it could be counter-productive. A subsequent output change cannot be allowed until ground bounce has died away, which could be longer than the extra delay introduced by limiting slew rate.

The solution lies in careful design and best use of the resources offered by the device manufacturer.

8.2.2 Signal reflections

Another cause of ringing on outputs is signal reflection. This is due to the signal tracks becoming transmission lines, a situation that arises as the delay time between an output and its load approaches the transition time (i.e. rise or fall time) of the signal.

Most device inputs have a high impedance; they are virtually open circuits compared with the impedance of the signal line. They therefore absorb no energy from the transmitted signal, which is reflected back to the signal source.

With short lines, the signal has not changed much when the reflection is received, so there is little disturbance. If the line is long enough for a significant change in signal level, the source will compensate for the discrepancy. This causes overshoot at the load because the signal and reflection are summed. In the case where the line delay equals the transition time, the overshoot equals the original signal size.

Continued reflections cause compensation in both directions; this appears as ringing at the load.

There are various schemes which may be tried to reduce ringing. Because the worst case is a low impedance with a fast edge driving a high impedance, just the situation when an FPGA drives a CMOS input, we can try increasing source impedance, reducing load impedance or slowing the edge. The first two require extra components to be added to the circuit, the third is possible with a low slew rate option on the FPGA output.

8.2.3 Comparison with standard parts

Even a designer who is new to FPGAs will probably have laid out boards containing standard devices such as microprocessors, peripherals, memory, glue logic and PLDs, and analogue and interface functions. What is so special about FPGAs then?

In short, the answer is probably not a lot. We have seen how the slew rate adjustment and output placement can affect ground bounce and ringing. The other fact to remember is that dedicated function devices have their I/Os preplaced and tailored to minimize these effects whereas the FPGA designer has to work it out for himself.

As well as ground bounce and ringing, this can affect problems like crosstalk. Very often, the place-and-route software will assign most, or all, of the I/Os unless the designer preplaces critical signals. The layout software will assign I/Os to maximize cell utilization and minimize internal routing, not to produce a 'sensible' I/O arrangement. The designer may well find himself with bus numbering in a haphazard sequence, for example, or input and output busses intermixed on the device pinning.

In order to connect the bus tracks on the circuit board to his FPGA, he may need to make extensive use of vias, cross-overs and circuitous paths. These are the things which he would normally avoid as they lead to problems like cross-talk, and signal skew, since approximately 10 cms of track can add 1 ns to a signal path.

Board layout should follow the same rules as other integrated circuits with multiple supply pins, but the designer should bear in mind the impact of I/O placement and drive capability from the onset of the FPGA design.

8.3 Conclusions

8.3.1 Past

The past situation regarding programmable logic has been summarized in Chapter 1. The 1980s was a decade of growth in programmable logic, both in terms of number of users and in terms of number of pieces used. The decade started with simple PALs, the 16 and 64 product term families, and the main applications were interfacing microprocessors to other LSI circuits, and replacing two or three MSI logic functions with a single chip.

By the end of the decade the 16V8 and 22V10 could, between them, replace any of the 20 or 24-pin PALs of ten years earlier, thanks to the introduction of macrocells. Some manufacturers had also introduced devices in larger packages, which could assimilate two or three 22V10s, thereby starting the upward trend in complexity which has characterized the 1990s.

In 1980, most designers could happily write logic equations for address decoders, but many were still designing in terms of standard logic families and then trying to convert these designs to logic equations to put them into programmable logic. Logic equation assemblers were the only design tool available to designers of programmable logic but, by the end of the decade, schematic capture and VHDL were in widespread use.

Alongside programmable logic, masked ASICs were also meeting a growing demand for customizable logic with the state of the art growing from 1000 gates to tens of thousands over a ten-year span. It was this growth which fuelled the development of the more automated design tools, which have now been incorporated into programmable logic design flows.

8.3.2 Present

The rest of this book has, I hope, covered the present situation. The state of the art in 1995 is an FPGA with up to 300 I/Os and 30 000 gates, although designers still use 16V8s. After all, the 74LS00 has not been made obsolete yet even though 7000 of them could fit into a single FPGA.

Most of the FPGAs currently being used in production still lie in the 1–5000 gate range, and one of the main uses of larger FPGAs is as a development tool for masked ASICs. This is probably the reason why design tools have become universal; the logic can now be designed completely independently of target architecture, process or performance. Only when a design has been proved does a target device have to be specified, and this can be an FPGA or masked ASIC.

Device-independent design has been the driving force behind VHDL, but most schematic capture packages also offer this facility. Both methods also support top down design techniques, where a design is specified in layers, or levels, which may be proved separately before being combined back into the top level.

An interesting device development has been the ability of some FPGAs to configure part of their logic area as memory. Trying to build RAM from logic elements is very space consuming and results in an inefficient use of resources. RAM-based FPGAs already have memory cells built into the chip so it makes sense to offer the user the choice of whether to use them as logic/interconnect defining elements, or pure memory in their own right. The introduction of the FLEX–10K family with a dedicated section for memory or logic magafunctions represents, perhaps, the first of a new generation of architectures which will allow even more integration.

8.3.3 Future

Progress in electronics is achieved by a mixture of innovative architecture and pure technological advance. The simplest measure of the latter is feature size; the width of a transistor gate or contact window size, for example. Of course, every time that feature size is halved the number of transistors per unit area increases fourfold. Thus, for a given die size, a 5000-gate array in 1 μm geometry becomes a 20 000-gate array in 0.5 μm geometry.

Although this is true for the internal logic functions, the components which interface with the outside world cannot change. They must still drive 50 pF capacitative loads and include bond pads to allow wires to connect the die with the lead frame. The periphery for a given I/O count will remain unchanged so, if this is minimized, a set area is available for logic. If this is just filled in one feature size, a process change will either leave a portion of the die unused, or cause an increase in the number of available gates.

This situation has arrived for masked ASICs, but not yet for FPGAs which use more area per gate than a sea-of-gates gate array. The FPGA uses more area with routing resources, programming circuitry and programmable switches. Thus, a designer who in the past had a direct choice between an ASIC and FPGA for his 5000-gate circuit, now has a choice between a 20 000-gate ASIC and a 5000-gate FPGA. FPGAs, then, are probably destined to take over the semi-custom market at an increasingly high level of gates.

Architecturally, the question remains as to what is going to be designed in a 10 000+ circuit. PLDs have traditionally been used for random logic replacement, but 10 000 gates is a lot of random logic. It seems increasingly likely that future FPGA designs will become more structured. Using VHDL as a design tool will aid this process as structured programming is more feasible with a software approach than a hardware method.

The other possible route has been shown by FLEX–10K, as we saw above. FPGAs of this complexity will be capable of implementing a complete system, including memory, processor and interface, so having areas of an FPGA dedicated to certain types of function would make sense. This does not necessarily mean that a particular processor would be hard wired into one area of the chip; it is more likely that one area could contain programmable structures capable of being wired as PLAs, registers and state machines, another area suited to pure memory and the whole embedded in random logic.

Whatever the detail of the future, there is no doubt that FPGAs will become more widespread, and in system programming will become a standard feature with design becoming practically indistinguishable from software design.

Appendix A Device manufacturers

1. Actel Corporation

USA 955 East Arques Avenue, Sunnyvale, California, CA94086. Tel: +1 408 739–1010
UK Intec 2, Unit 22 Wade Road, Basingstoke, Hampshire, RG24 8NE. Tel: +44 (0)1256 29209
Germany Bahnhofstrasse 15, 85375 Neufahrn. Tel: +49 (0)8165–66101

Device range: ACT1, ACT2, ACT3 antifuse FPGAs – Section 6.1

Design support: Designer and Designer Advantage may be supplied bundled with Viewlogic PRO-series schematic capture and simulation, or with the equivalent OrCAD products; Cadence and Mentor can also be provided for workstations. Most other CAD companies are supported in Actel's Industry Alliance Program. Schematic netlists created with third-party CAD are converted to the Actel format through an EDIF interface.

ACTmap is an interface between text-based design methods, PALASM and VHDL, and schematics. Blocks of logic defined textually are given a graphical symbol for incorporation with other logic symbols in the schematic entry phase. ACTmap interprets the text-based logic definition into Actel logic modules.

Designer supports designs up to 2500 gates, Designer Advantage up to 10 000 gates. They use the Actel netlist to place-and-route the design. Critical paths may be defined in advance to prompt the autorouter to minimize delay for critical signals. ChipView and ChipEdit allow the designer to monitor module placement and, after layout, make manual adjustments to try to improve performance.

The Actel timer will predict delay times between specified nodes. It also provides the back-annotated network file for post-layout simulation and timing analysis.

Programming support: The Activator 2S (single station) and Activator2 (four station) provide programming support for all Actel FPGAs. Each family/package combination needs a separate adaptor.

The Actionprobe diagnostic tools allow internal nodes of any Actel device to be monitored in-system, provided that the probe pins are not used for logic signals in normal operation.

Data I/O also support ACT parts on the Unisite programmer.

2. Advanced Micro Devices Inc.

USA One AMD Place, PO Box 3453, Sunnyvale, California CA94088-3453. Tel: +1 408 732-2400/+1 800 538-8450
UK AMD House, Goldsworth Road, Woking, Surrey, GU21 1JT. Tel: +44 (0)1483 740440/+44 (0)1925 830380 (Warrington office). Fax: +44 (0)1483 756196/+44 (0)1925 830204
Germany Rosenheimer Strasse 143B, 8000 Munchen 80. Tel: +49 (0)89-450530/+49 (0)6172-24061 (Bad Homburg office). Fax: +49 (0)89-406490/+49 (0)6172-23195
France Tel: +33 (0)1 49-75-10-10 (Paris office). Fax: +33 (0)1 49-75-10-13
Japan Tel: +81 (0)3 3346-7550 (Tokyo office)/+81 (0)6 243-3250 (Osaka office). Fax: +81 (0)3 3346-7848/+81 (0)6 243-3253

Device range: MACH1, 2, 3 and 4 ranges of LSI PALs – Section 4.1

Design support: PALASM4 is a text-based design entry system for Boolean and State equations. PALASM4 supports only the MACH1 and -2 families and is unlikely to be developed further. It includes an event-driven simulator and can accept schematic netlists from major CAD vendors.

MACHXL is based on PALASM4 and offers similar support. New devices will be added to this product, rather than PALASM4.

AMD also supply dedicated versions of Design Center, ABEL and PROdeveloper (Viewlogic). All three support Schematic, Boolean and VHDL entry, functional and timing simulation.

MACH fitters are available for most of the standard CAD packages through the FusionPLD program. This ensures that designers who already use a CAD package can add MACH capability with minimum overhead, and maximum confidence. AMD approve and certify all MACH fitters before they are released for customer use.

Programming support: AMD have approved seven programmer manufacturers through the FusionPLD program, although most programmers will support the MACH families, often with the use of additional socket adaptors.

3. Altera Corporation

USA 2610 Orchard Parkway, San Jose, California, CA95134–2020. Tel: +1 408 894–7000 Fax: +1 408 944–0952
UK Solar House, Globe Park, Fieldhouse Lane, Marlow, Bucks, SL7 1TB (European Headquarters). Tel: +44 (0)1628 488800 Fax: +44 (0)1628 890078
Japan Dai Tokyo Kasai Shinjuku Bldg., 3–25–3 Yoyogi, Shibuya-ku, Tokyo 151. Tel: +81 (0)3 3375–2281 Fax: +81 (0)3 3375–2287

Device ranges: MAX5000, 7000, 9000 LSI PAL-type CPLDs – Section 4.2; FLASHlogic LSI PAL-type CPLDs – Section 4.4; FLEX8000/10K RAM-based FPGAs – Section 5.3.

Design support: All Altera ranges are supported by the proprietary MAX+PLUS II design system. Design entry is device independent and can be by schematic capture or a variety of text-based entry methods; VHDL, Verilog HDL or Altera's own AHDL may all be used. Waveform design entry is also possible. A synthesis algorithm will convert the waveforms into logic streams, generating the relationship between them and converting this into logic structures.

The Floorplan Editor assists pin assignment and allows the designer to pre-place logic modules to force the compiler into a particular routing pattern. Alternatively, a design can be compiled cold and modified after compilation to adjust timings. The pre-placement can also be accomplished with a textual file.

Part of the compilation process is a pre-processing phase which analyses the design data for redundant logic and performs minimization on the stripped out logic file. Desired timing parameters can be included to guide the compiler into the appropriate placement for critical paths. As well as fitting the logic into the target, the compiler produces output files for programming, simulation and post-layout timing analysis.

Multi-device partitioning is available for designs which are too large for a single device. The fitter will use the smallest number of parts from one family and minimize the interconnections between parts. User intervention is possible at all stages to influence the compiler in its partitioning and fitting phases.

Altera have established partnerships with third-party CAD suppliers through its ACCESS (Altera Commitment to Cooperative Engineering

Solutions) program. The MAX+PLUS II compiler accepts the EDIF files from the CAD software, and processes it as the normal back end function described above.

Programming support: Altera can provide an MPU (master programming unit) which supports all those devices which are programmed out of circuit. A FLEX Download Cable can be attached to the MPU or a BitBlaster connected to the PC serial port for in-circuit configuration of RAM-based FPGAs.

Many third-party programmers offer programming support for the MAX and FLASHlogic families.

4. Atmel

USA 2125 O'Nel Drive, San Jose, California, CA95131. Tel: +1 408 441−0311 Fax: +1 408 436−4200

UK Coliseum Business Centre, Riverside Way, Camberley, Surrey, GU15 3YL. Tel: +44 (0)1276 686677 Fax: +44 (0)1276 686697

France 55 Avenue Diderot, 94100 St. Maur des Fosses, Paris. Tel: +33 (0)1 48−85−55−22 Fax: +33 (0)1 48−85−55−96

Germany Ginnheimer Strasse 45, 60487 Frankfurt 90. Tel: +49 (0)69−707 5910 Fax: +49 (0)69−707 5912

Japan NT Building, 1−9−12, Uchikanda, Chiyoda-Ku, Tokyo 101. Tel: +81 (0)3−5259−0211 Fax: +81 (0)3−5259−0217

Device range: AT6000 series of RAM-based FPGAs − Section 5.2

Design support: The AT6000 design system is based on the Viewlogic schematic capture and simulation package. Options include an add-on utility for existing Viewlogic users, a capture only package and the full schematic capture and simulation suite. The Atmel Design Manager controls the design flow and includes the place-and-route software, with or without intervention, timing analyzer, design rule check and bit stream generator. A common database approach streamlines the design flow while maintaining design integrity.

Programming support: Because AT6000 is a RAM-based technology, devices are usually programmed in-circuit in production. A prototype kit allows a device to be programmed in standalone mode for checking or building into a prototype system.

5. AT&T Microelectronics

USA 555 Union Boulevard, Room 21Q-133BA, Allentown, PA 18103. Tel: +1 800 372−2447 Fax: +1 610−712−4106

UK Powell Duffryn House, London Road, Bracknell, Berkshire, RG12 2AQ. Tel: +44 (0)1344 487111 Fax: +44 (0)1344 485735

France Tour Horizon, 52 Quai de Dion Bouton, F-92800 Puteau Cedex, Paris. Tel: +33 (0)1 48–85–55–22 Fax: +33 (0)1 48–85–55–96
Germany (Berlin office) Tel: +49 (0)30–63925370 Fax: +49 (0)30–6643225 (Munich office) Tel: +49 (0)89–95086–0 Fax: +49 (0)89–95086–333
Japan (Osaka office) Tel: +81 (0)6–945–6515 Fax: 81 (0)6–945–6530 (Tokyo office) Tel: +81 (0)3–5421–1600 Fax: +81 (0)3–5421–1700

Device range: ORCA 2C series of RAM–based FPGAs – Section 5.4; Second source of Xilinx 3000 series – Section 5.1.

Design support: AT&T offer an 'FPGA foundry' for place-and-route and generating the programming bitstream. It interfaces to standard front–end design entry tools, particularly Mentor and Viewlogic, and various VHDL synthesis tools. The foundry Timing Wizard tool attempts to ensure that the finished circuit meets specified timing parameters included in the designer's preference file.

Programming support: Being RAM-based, ORCA 2C FPGAs are programmed in-circuit with data stored in an adjacent EPROM.

6. Cypress Semiconductor

USA 3901 North First Street, San Jose, California, CA95134. Tel: +1 408 943–2600 Fax: +1 408 943–2741
UK 3 Blackhorse Lane, Hitchin, Hertfordshire, SG4 9EE. Tel: +44 (0)1462 420566/+44 (0)1614 282208 (Manchester office). Fax: +44 (0)1462 421969/+44 (0)1614 280764
France Miniparc Batiment no.8, Avenue des Andes, Z.A. de Courtaboeuf, 91952 Les Ulis Cedex. Tel: +33 (0)1 69–07–55–46 Fax: +33 (0)1 69–07–55–71
Germany Munchner Strasse 15A, W–8011 Zorneding. Tel: +49 (0)81–062855/+49 (0)4193–77217 (Northern office). Fax: +49 (0)81–0620087/+49 (0)4193–78259
Japan Fuchi-Minami Building, 2F 10–3, 1-Chome, Fuchu-machi, Fuchu-shi, Tokyo 183. Tel: +81 (0)423–698211 Fax: +81 (0)423–698210

Device ranges: CY7C340 series are second source to the MAX5000 series – Section 4.2; CY7C370 series are LSI PAL-type parts – Section 4.7; pASIC380 family are second source to the QL FPGAs – Section 6.2.

Design support: Warp2 is a VHDL compiler for the CY7C340s and CY7C370s (CPLD ranges). It accepts textual input in VHDL, standard Boolean or state equation format, and synthesizes the optionally mixed input into a logical representation. This is optimized to a minimal logic form, and fitted to the target device, resulting in a programming file.

Waveform simulation may be performed with the Nova simulator, which will produce a test vector file to add to the programming file.

Warp3 is the universal development system for all Cypress PLDs and FPGAs. It has a combined VHDL/schematic front end, accepting either graphics or text; the schematic capture package is the industry standard Viewdraw. In mixed mode, a symbol editor generates symbols for blocks of VHDL which can then be included with the general schematic capture environment. Schematics are converted to VHDL before compilation.

Functional simulation may be combined with a VHDL debugger in which the result of each step of VHDL code is displayed graphically.

Device fitting or, for FPGAs, place-and-route is preceded by a logic optimization phase in which the Boolean logic is minimized and registered logic is fitted to the most appropriate flip-flop type or state machine format. Pin assignment may be manual or automatic, but place-and-route needs no manual intervention for optimum results.

After place-and-route, timing delays may be back-annotated to the simulator for prediction of device performance.

Programming support: Cypress offer the Impulse3 programmer, which covers all their PLD/FPGA ranges. Most third-party programmers also support the Cypress families.

7. ICT Inc.

USA 2123 Ringwood Avenue, San Jose, California, CA95131. Tel: +1 408 434–0678 Fax: +1 408 432–0815

UK representative Siretta Microelectronics Ltd., Tekelec House, Back Lane, Spencers Wood, Reading, Berks, RG7 1PD. Tel: +44 (0)1734 258080 Fax: +44 (0)1734 258070

France representative Microel. Tel: +33 (0)1 69–07–08–24

Germany representatives Tekelec. Tel: +49 (0)89–51640; USE. Tel: +49 (0)89–339292

Japan representatives Sanei Electronic Industry Co. Tel: +81 (0)426 447338; Tokyo Denshi Hanbai Co. Tel: +81 (0)333 483401

Device range: PA7140 LSI sequencer – Section 4.6

Design support: PLACE is supplied free by ICT, and covers all their product ranges. The editor allows logic definition by Boolean equations, state equations or truth tables. The cell configurations may be defined and viewed graphically, allowing the designer to tailor the device architecture to particular requirements.

Compilation includes logic minimization, ensuring a good fit for the specified design. A JEDEC programming file is produced by the compiler.

The simulator displays results in graphical format, allowing good visibility of the results, and includes a graphical editor. The simulator is event driven; that is, it responds only to changes in input levels and does not include timing information. Simulation results can be added to the programming file as test vectors.

Programming support: The ICT PDS-3 programmer supports the PA7140. Most third-party programmers may also be used for programming this device.

8. Lattice Semiconductor Corp.

USA 5555 Northeast Moore Ct., Hilsboro, Oregon 97124. Tel: +1 503 681-0118 Fax: +1 503 681-3037
UK Castle Hill House, Castle Hill, Windsor, Berkshire SL4 1PD. Tel: +44 (0)1753 830842 Fax: +44 (0)1753 833457
France Les Algorithmes, Batiment Homere, 91 190 St. Aubin. Tel: +33 (0)1 41-14-83-00 Fax: +33 (0)1 42-53-97-31
Germany Hanns-Braun-Strasse 50, 85375 Neufahrn-bei-Munchen. Tel: +49 (0)8165-9516-0 Fax: +49 (0)8165-9516-33
Japan N Bld 9F, 2-28-3, Higashi-Nihonbashi, Chuo-ku, Tokyo 103. Tel: +81 (0)3-5820-3533 Fax: +81 (0)3-5820-3531

Device range: (is)pLSI 1000, 2000 and 3000 series of LSI PAL-type devices – Section 4.5

Design support: Lattice offer their PDS system as a standalone development system with Boolean entry and assignment of macros into GLBs. Logic minimization is performed after logic entry to reduce the number of product terms required for each output function.

The compiler can be run in two modes. Fast route offers a quick solution which is not necessarily optimized for resource utilization. Strong route offers a slower, but more thorough routing scheme, which gives the best solution possible. It will also permit small design changes to be made without prejudicing the existing pin-out.

Functional and post-layout simulation may be performed with optional third-party tools, such as Viewsim.

Lattice also provides back-end support to most major third-party CAD providers. This enables schematic capture or VHDL entry methods to be interfaced to the PDS system.

Programming support: pLSI devices must be programmed in standalone mode. Several programmer manufacturers offer support for Lattice LSI parts. The isP Engineering Kit provides support for the in-system programmed parts. For development purposes a download cable with Programming Module can be supplied. This takes the bitstream produced by the PDS software, and transfers it to the target device. It may also be used for production, although it is more usual to provide in-system programming at board level.

9. Motorola

USA Logic Integrated Circuits Division, PO Box 20906, Phoenix, Arizona 85036. Tel/fax: call local sales office
UK Motorola Ltd., Fairfax House, 69 Buckingham Street, Aylesbury, Bucks, HP20 2NF. Tel: +44 (0)1296 395252 Fax: +44 (0)1296 21999

Device range: MPA1000 RAM-based FPGAs – Section 5.4

Design support: Based on third-party tools such as Viewlogic and Synopsys for design entry and simulation. The NeoCAD FPGA Foundry is used for layout and timing analysis.

Programming support: Being a RAM-based FPGA, configuration data is stored in a separate on-board store. Motorola offer the MCP37LV128 128k serial EPROM for direct interface to the MPA1000 series.

10. Philips Semiconductors

USA 811 East Arques Avenue, Sunnyvale, California CA 94088–3409. Tel: +1 800 234–7381 Fax: +1 708 296–8556
UK 276 Bath Road, Hayes, Middlesex, UB3 5BX. Tel: +44 (0)181–730 5000 Fax: +44 (0)181–754 8421
France 4 Rue du Port-aux-Vins, BP317, 92156 Suresnes Cedex. Tel: +33 (0)1 40–99–61–61 Fax: +33 (0)1 40–99–64–27
Germany PO Box 10 63 23, 20095 Hamburg. Tel: +49 (0)40 3296–0 Fax: +49 (0)40 3296–213
Japan Philips Bldg.13–37, Kohnan 2-chome, Minato-ku, Tokyo 108. Tel: +81 (0)3 3740 5101 Fax: +81 (0)3 3740 0570

Device range: LSI sequencers PML2552/2852 – Section 4.6

Design support: SNAP (Synthesis Netlist Analysis Program) is a standalone program accepting design information in Boolean, state equation or truth table formats. It will also interface to schematic capture and VHDL packages with the EDIF format.

Simulation may be employed both before and after compilation to check both the design input and device implementation. The results of simulation can be incorporated into the JEDEC programming file as test vectors. A fault simulator provides test coverage information allowing the designer to add test patterns until the device is fully tested.

Programmer support: PML parts are supported by Data I/O, Strebor and Basic Computer Systems AG.

11. QuickLogic Corporation

USA 2933 Bunker Hill Lane, Santa Clara, California, CA95054. Tel: +1 408 987–2000 Fax: +1 408 987–2012
UK representative Abacus Electronics Ltd., Abacus House, Bone Lane, Newbury, Berkshire, RG14 5SF. Tel: +44 (0)1635 36222 Fax: +44 (0)1635 38670
France representative MISIL Technologies, 2 Rue de la Couture, Silic 301, 94588 Rungis Cedex. Tel: +33 (0)1 45–60–00–21 Fax: +33 (0)1 45–60–01–86
Germany representative Scantec Mikroelektronik GmbH, Behringstrasse 10, D–82152 Planegg. Tel: +49 (0)89–8598–021 Fax:+49 (0)89–8576–574
Japan 2–36–12 Yoga, Setagaya-ku, Tokyo 158. Tel: +81 (0)3–5716–5930 Fax: +81 (0)3–5716–5931

Device range: pASIC (QL series) antifuse FPGAs – Section 6.2

Design support: QuickLogic offer an interface to a variety of standard CAD tools, centred on the SpDE (Seamless pASIC Design Environment) design manager. Design entry by Boolean equations, HDLs, schematic capture, or a mixture of all three, is converted to an EDIF standard netlist, or to 'QDIF', either of which is accepted by SpDE. Functional simulation at this stage is achieved by using the EDIF data and a standard simulator.

SpDE performs automatic place-and-route to produce a final chip layout. This layout may be viewed and, in conjunction with the delay modeller, any path highlighted and its delay displayed. The net will also be highlighted in a schematic display, if a capture tool has been used. Any delays which are too slow can be tagged and place-and-route re-run to achieve a specified delay.

The delay modeller also generates the timing netlist for post-layout simulation, again using a third-party tool.

The programming file with test vectors is the final product of SpDE.

Programming support: A dedicated programmer is available from Quick-Logic, using adaptors to cater for the different package types. pASIC devices are also supported by Data I/O and SMS.

12. Xilinx Inc.

USA 2100 Logic Drive, San Jose, California, CA95124. Tel: +1 408 559–7778   Fax: +1 408 559–7114
UK Suite 1B, Cobb House, Oyster Lane, Byfleet, Surrey, KT14 7DU. Tel: +44 (0)1932 349401 Fax: +44 (0)1932 349499
France Z.I. de la Bonde-Batiment B, 1 bis, Rue Marcel Paul, 91742 Massey Cedex. Tel: +33 (0)1 60–13–34–34 Fax: +33 (0)1–60–13–04–17
Germany Dorfstrasse 1, 85609 Ascheim. Tel: +49 (0)89–904–5024 Fax: +49 (0)89–904–4748
Japan Daini-Nagaoka Building 2F, 2–8–5, Hatchobori, Chuo-ku, Tokyo 104. Tel: +81 (0)3–3297–9191 Fax: +81 (0)3–3297–9189

Device ranges: XC7200A and XC7300 LSI PAL-type devices – Section 4.3; XC2000, 3000, 4000 and 5000 RAM-based FPGAs – Section 5.1; XC8100 antifuse FPGAs – Section 6.3

Design support: The XACT 5.0 design system uses a common library of components and design flow for all Xilinx families, except XC8100. Interfaces are available for most of the major CAD suppliers in both schematic capture and VHDL entry formats. Standard text-based methods, using Boolean equations or state equations are directly supported by the compiler for conversion to a netlist. The design process is managed by the Xilinx Design Manager (XDM).

Another option is X-BLOX, a graphics-based synthesis tool which may be interfaced to XACT. Logic modules are defined as behavioural blocks but connected as graphical symbols. This offers designers the power of an HDL with the convenience and visibility of graphics.

The XC8100 family uses a system similar to XACT, but with a different component library to take account of the new architecture. Initially this XC8100 Development System (XDS) is supplied only with synthesis tools, but later versions based on standard CAD tools should include schematic capture.

Both systems make use of the third-party simulators for both functional and post-layout simulation.

Design implementation may be fully automatic or under varying degrees of user intervention. The designer can specify the floorplan of an FPGA for up to 100% of the cells, and pre-define routing to allow critical timing paths to benefit from the minimum interconnect lengths.

Because the component libraries are common across all families, migration between parts is easy in the event of logic content or performance in the original target proving inadequate.

Programming support: The XC7200A, XC7300 and XC8100 families need programming before assembly into boards. Xilinx can supply a standalone programmer, or most third-party manufacturers can also offer support.

The RAM-based FPGAs are in-circuit programmable. A parallel download cable and XChecker cables provide development support. The parallel cable offers a high speed download of bitstream data into the target device on-board. The XChecker works from the serial port, but also provides a readback capability which is capable of accessing internal nodes.

In production, data must be loaded into the FPGA at power-up. This may be from an EPROM or from a Xilinx serial PROM, which has the advantage of occupying much less board space, and interfaces directly with only a single control pin.

Appendix B CAD and programmer suppliers

In this appendix, the suppliers are listed with the names of their products. Product descriptions are best obtained directly from the supplier. Full colour brochures will give a good idea of the type of display and features offered; very often a demonstration disk is also available to give a prospective buyer an indication of how the product will appear on his own system.

Accel Technologies

USA 6825 Flanders Drive, San Diego, California, CA92121. Tel: +1 619 554–1000 Fax: +1 619 554–1019

Product range: TangoPLD – design software.

Advin Systems Inc.

USA 1050–L Duane Avenue, Sunnyvale, California, CA94086. Tel: +1 408 243–7000 Fax: +1 408 736–2503
UK agent Quarndon Electronics Ltd., Slack Lane, Derby, DE3 3ED. Tel: +44 (0)1332 332651 Fax: +44 (0)1332 360922
France Antycip, 18 rue du Perouzet, 95100 Argenteuil. Tel: +33 (0)1 39–61–14–14 Fax: +33 (0)1 30–76–29–73
Germany Lascar Electronics GmbH, Vordere Kirchstrasse 4, D–7241 Eutingen–2. Tel: +49 (0)7–459–1271 Fax: +49 (0(7–459–2471

Product range: Pilot–U84 universal device programmer.

Basic Computer Systems A.G.

Austria Wolfgang-Pauli-Gasse, A–1140 Klagenfurt-Auhof. Tel: +43 (0)222–9736360 Fax: +43 (0)222–975915

Product range: UP2000 device programmer.

BP Microsystems Inc.

USA 1000 North Post Oak Road, Suite 225, Houston, Texas, TX77055. Tel: +1 713 688–4600 Fax: +1 713 688–0902
UK agent Mutek Ltd., Farleigh House, Frome Road, Bradford on Avon, Wiltshire,. Tel: +44 (0)1225 866501 Fax: +44 (0)1225 865083
France Emulations, Antelia 4 Burospace, Chemin de Gizy, 91751 Bievres Cedex. Tel: +33 (0)1 69–41–28–01 Fax: +33 (0)1 60–19–29–50
Germany API Electronik, Lorenz-Braren Strasse 32, 8062 Markt Indersdorf. Tel: +49 (0)81 367092 Fax: +49 (0)81 367398

Product range: BP1200, CP1128, PLD1100 device programmers.

Cadence Design Systems Inc.

USA 555 River Oaks Parkway, San Jose, California, CA95134. Tel: +1 408 943–1234 Fax: +1 408 943 0513
UK Cadence Design Systems Ltd., Bagshot Road, Bracknell, Berkshire, RG12 3PH. Tel: +44 (0)1344 360333 Fax: +44 (0)1344 869660
Germany Kirchheim. Tel: +49 (0)89 9050–910 Fax: +49 (0)89 9044–925
Japan Tel: +81 (0)33 497–4460 Fax: +81 (0)33 497–4469

Product range: Design Framework II – CAD suite including VHDL, schematic capture, logic synthesis and simulation, running on workstations.

Data I/O Corp.

USA 10525 Willows Road N.E., PO Box 97046, Redmond, WA98073–9746. Tel: +1 206 881–6444 Fax: +1 206 882–1043
France Data I/O Europe, 106 rue des Freres Farman, BP 328, 78533 Buc Cedex. Tel: +33 (0)1 39–56–46–57
UK Data I/O Ltd., 660 Eskdale Road, Winnersh, Wokingham, Berkshire, RG11 5TS. Tel: +44 (0)1734 440011 Fax: +44 (0)1734 448700
Germany Data I/O GmbH, Lochhamer Schlag 5, 82166 Graefelfing. Tel: +49 (0)89 858580

Japan Data I/O Japan, Sumitomoseimei Higashishinbashi, Bldg. 8F, 2-1-7, Higashi-Shinbashi, Minato-ku, Tokyo 105. Tel: +81 (0)33 432-6991

Product range: ABEL6 – HDL, Boolean and state equation design entry and compilation package for CPLDs; Synario – universal FPGA design system with mixed mode entry, bundled place and route, and simulation; Unisite – universal device programmer; Chiplab – development programmer.

Elan Digital Systems Ltd.

UK Elan House, Little Park Farm Road, Segensworth West, Fareham, Hants, PO15 5SJ. Tel: +44 (0)1489 5797999 Fax: +44 (0)1489 577516
USA Elan Systems Inc., 365 Woodview Avenue, Suite 700, Morgan Hill, California, CA95037. Tel: +1 408 778-7267 Fax: +1 408 778-2597

Product range: Model 6000 Programming System – universal device programmer.

Intergraph Corporation

USA Huntsville, Alabama 35894-00001. Tel: +1 205 730-2000
UK Intergraph (UK) Ltd., Delta Business Park, Great Western Way, Swindon, Wiltshire, SN5 7XP. Tel: +44 (0)1793 619999 Fax: +44 (0)1793 618508
Holland Intergraph Europe Inc., PO Box 333, Hoofddorp, Netherlands. Tel: +31 (0)2503-66333
Hong Kong Intergraph Asia-Pacific Ltd., Room 903, Harcourt House, 39 Gloucester Road. Tel: +852 8661966

Product range: VeriBest Design System/TD-1 workstation – integrated software/hardware package for a complete design environment, although VeriBest also runs on PC. VHDL/Verilog HDL input may be mixed with schematic capture on the ACEPlus Designer. The VeriBest simulator is Verilog compatible, with results displayed graphically using VeriScope.

ISDATA GmbH

Germany Haid-und-Neu-Strasse 7, 7500 Karlsruhe 1. Tel: +49 (0)721-693092 Fax: +49 (0)721-174263
USA ISDATA Inc., 800 Airport Road, Monterey, California, CA93940. Tel: +1 408 373-7359 Fax: +1 408 373-3622

Product range: LOG/iC – design entry system supporting VHDL, other text-based modes and schematic/state machine graphics. LOG/iC also includes a functional simulator.

Logical Devices Inc.

USA 692 South Military Trail, Deerfield Beach, Florida, FL33442. Tel: +1 305 428–6868 Fax: +1 305 428–1181

Product range: CUPL – text-based design entry system, including functional simulation. ALLPRO – Universal device programmer.

Mentor Graphics

USA 8005 S.W. Boeckman Road, Wilsonville, OR97070. Tel: +1 503 685–7000 Fax: +1 503 685–1204
UK Mentor Graphics (UK) Ltd., Rivergate, Newbury Business Park, London Road, Newbury, Berks, RG13 2QB. Tel: +44 (0)1635 811411 Fax: +44 (0)1635 811108

Product range: ASIC Design Suite – workstation-based ASIC/FPGA toolset for integrated design using schematic capture or VHDL entry.

Minc Inc.

USA 6755 Earl drive, Colorado Springs, Colorado, CO80918. Tel: +1 719 590–1155 Fax: +1 719 590–7330

Product range: PGADesigner – mixed entry mode logic synthesizer with functional simulation.

NeoCAD Inc.

USA 2585 Central Avenue, Boulder, Colorado, CO8030. Tel: +1 303 442–9121 Fax: +1 303 442–9124

Product range: FPGA Foundry – a complete FPGA design system from design capture, through synthesis and simulation to timing driven layout and analysis.

OrCAD Systems Corp.

USA 9300 SW Nimbus Avenue, Beaverton, OR97005. Tel: +1 503 671–9500 Fax: +1 503 671–9501

France OrCAD Europe, 96 rue St. Charles, 75015 Paris. Tel: +33 (0)1 45-75-50-00. Fax: +33 (0)1 45-77-82-89

Product range: A suite of design tools comprising: SDT386+ – schematic capture design entry tool; PLD386+ – logic synthesizer/compiler for CPLDs and FPGAs; VST386+ – functional and timing simulator.

PREP Corp.

USA 504 Nino Avenue, Los Gatos, CA95032. Tel: +1 408 356-2169 Fax: +1 408 356-0195

Product range: Benchmark data for comparison of FPGAs. Information is available on a World Wide Web Homepage at *http://www.prep.org* free of charge.

SMS Micro Systems GmbH

Germany IM Grund 15, D-7988 Wangen. Tel: +49 (0)7522-5018 Fax: +49 (0)7522-8929
USA SMS North America Inc., 16522 N.E. 135th Place, Redmond, WA98052. Tel: +1 206 883-8447 Fax: +1 206 883-8601

Product range: Sprint Expert/Plus – universal device programmer.

Stag Programmers Ltd.

UK Silver Court, Watchmead, Welwyn Garden City, Herts AL7 1BR. Tel: +44 (0)1707 332148 Fax: +44 (0)1707 371603
USA Stag Microsystems Inc., 1600 Wyatt Drive, Santa Clara, California, CA95054. Tel: +1 408 988-1118 Fax: +1 408 988-1232

Product range: Eclipse and Quasar – universal device programmers. Stag also act as UK agents for Logical Devices CUPL and ISDATA LOG/ic software.

Strebor Data Communications

USA 1008 N. Nob Hill, American Fork, Utah 84003. Tel: +1 801 756-3605

Product range: PLP-S1/S1A programmer for Philips PML family.

Synopsis Inc.

USA 700 East Middlefield Road, Mountain View, California, CA94043. Tel: +1 415 962–5000 Fax: +1 415 965–8637
UK Synopsis (Northern Europe) Ltd., The Imperium, Worton Grange, Reading, Berkshire, RG2 0TD. Tel: +44 (0)1734 313822 Fax: +44 (0)1734 750081
Germany Synopsys Europe, Suskindstrasse 4, D–81929 Munich. Tel: +49 (0)89–9939–1230 Fax: +49 (0)89–9939–1232
Japan Nihon Synopsys, Dai-Tokyokasai Shinjuku Building 6F, 3–25–3 Yoyogi, Shibuya-ku, Tokyo 151. Tel: +81 (0)3–5351–5300 Fax: +81 03–5351–5451

Product range: A complete set of VHDL design tools including, modelling, simulation, synthesis at behavioural or RTL levels, design for test and vendor interface.

System General

USA 510 South Park Victoria Drive, PQ Box 361898, Milpitas, California 95036–1898. Tel: +1 408 263–6667

Product range: SGUP–85/85A and TURPRO–1 programmers.

Viewlogic Systems Inc.

USA 293 Boston Post Road West, Marlboro, MA01752–4615. Tel: +1 508 480–0881 Fax: +1 508 480–0882
UK Viewlogic Systems Ltd., Berkshire Court, Bracknell, Berkshire, RG12 1RE. Tel: +44 (0)1344 303737 Fax: +44 (0)1344 340747
Japan Viewlogic Japan, King Hills Hakusan 9F, 2–25–10, Hakusan, Bunkyo-ku, Tokyo 112. Tel: +81 (0)35–689–7575 Fax:+81 (0)35–689–7574

Product range: Design suite for schematic capture/VHDL and simulation: PROcapture/Viewdraw – schematic capture program; PROsynthesis – VHDL design program; PROsim/Viewsim – VHDL-based timing simulator; PROwave/Viewwave – graphical simulator interface.

Appendix C References

Bostock, Geoff (1993) *Programmable Logic Handbook*. 2nd. Ed., Oxford, Butterworth Heinemann

Ott, Douglas E., Wilderotter, Thomas J., (1994) *A Designers Guide to VHDL Synthesis*. Kluwer Academic Publications

Data Sheets and Application Notes published by the device and CAD suppliers listed in Appendices A and B. Most of the device manufacturers' data books contain a large selection of application notes and include references, too numerous to list here, to published papers on most aspects of FPGA design and application. Many manufacturers also provide bulletin board services for direct access to device data and applications.

Appendix D   Trade Marks

The following trade marks and registered trade marks were used in this book:

ORCA attributed to AT&T Microelectronics
TangoPLD attributed to Accel Technologies Inc.
ACT, Actionprobe, Designer, Designer Advantage, ACTmap, Activator attributed to Actel Corporation
AMD, PAL, PALASM, MACH, MACHXL, FusionPLD attributed to Advanced Micro Devices
MAX, FastTrack, MAX+PLUS, FLASHlogic, FLEXlogic, FLEX, BitBlaster attributed to Altera Corporation
Cache Logic attributed to Atmel Corporation
Verilog attributed to Cadence Design Systems Inc.
FLASH370, Warp2, Warp3, Impulse3 attributed to Cypress Semiconductor
ABEL, Unisite, Chiplab, Synario attributed to Data I/O Corporation
PLACE attributed to ICT Inc.
LOG/iC attributed to ISDATA GmbH
Intergraph, VeriBest, ACEPlus attributed to Intergraph Corporation
GAL, ispLSI, pLSI, pDS+ attributed to Lattice Semiconductor Corporation
CUPL, ALLPRO attributed to Logical Devices Inc.
Mentor attributed to Mentor Graphics Inc.
PGADesigner attributed to Minc Inc.
NeoCAD, FPGA Foundry attributed to NeoCAD Inc.
OrCAD, PLD386+, SDT386+, VST386+ attributed to OrCAD Systems Corporation
PML, SNAP attributed to Philips Semiconductors

PREP attributed to Programmable Electronics Performance Corporation
QuickLogic, ViaLink, SpDE, pASIC attributed to QuickLogic Corporation
SPRINT Plus, SPRINT EXPERT attributed to SMS Micro Systems GmbH
Eclipse, Quasar attributed to Stag Programmers Ltd.
Synopsis attributed to Synopsis Inc.
Viewlogic, Viewdraw, Viewsim, Viewwave, PRO-series, PROcapture, PROsynthesis, PROsim, PROwave attributed to Viewlogic Systems
LCA, Fastclock, VersaBlock, VersaRing, XACT, MicroVia, Configurable Logic Cell, X-BLOX, XDM, XDS, XChecker attributed to Xilinx Corporation

Glossary

AC	Alternating current
ASIC	Application-specific integrated circuit
CE	Count enable
CFB	Configurable function block
CLB	Configurable logic block
CLC	Configurable logic cells
CLK	Clock
CMOS	Complementary metal/oxide semiconductor
Connects	Metallized horizontal interconnections between solid-state components in an IC array
CPLD	Complex programmable logic device
D/E	Data enable
DC	Direct current
EAB	Embedded array block
EEPROM	Electrically-erasable programmable read-only memory
EPROM	Erasable programmable read-only memory
FFB	Fast function block
FIFO	First in, first out
FPGA	Field programmable gate array
FPLA	Field programmable logic array
FPLS	Field programmable logic sequencer
GAL	Generic array logic
GLP	General logic block
GND	Ground
GRP	Global routing pool
HDFB	High density function block
HDL	Hardware description language

I/O	Input/output
IC	Integrated circuit
JTAG	Joint Test Action Group
K-map	Karnaugh map
LAB	Logic array block
LCA	Logic cell array
LE	Logic element
LE	Logic element
LSI	Large scale integration
LUT	Look-up table
MACH	Macro array CMOS high density
MAX	Multiple array matrix
MPU	Master programming unit
MSI	Medium scale integration
MTBF	Mean time between failures
NRE	Non-recurring engineering
ORCA	Optimized reconfigurable cell array
ORP	Output routing pool
PAL	Programmable array logic
PCB	Printed circuit board
PFU	Programmable function unit
PIA	Programmable interconnection array
PIM	Programmable interconnect matrix
PIN	Personal identification number
PIP	Programmable interconnection points
PLD	Programmable logic device
RAM	Random-access memory
SRAM	Static random-access memory
SSI	Small scale integration
TAP	Test access point
TMS	Test mode select
TTL	Transistor-transistor logic
UART	Universal asynchronous receiver/transmitter
UIM	Universal interconnect matrix
UV	Ultra-violet (light)
VHDL	VHSIC hardware description language
VHSIC	Very high speed integrated circuit
Vias	Vertical links between connects
VLSI	Very large scale integration

Index

3.3 volt interface
  FLASHlogic, 92
  FLEX10k family, 125
  FLEX8000 family, 125
  MPA1000 family, 129
  XC2000L family, 113
  XC3000L family, 115
  XC7300 family, 89
4000 series CMOS, 1

ABEL, 47, 202
Accel Technologies, 200
ACCESS, 191
ACEPlus Designer, 202
ACT1 family, 130
ACT2 family, 133
ACT3 family, 135
Actel, 130, 189
Actionprobe, 137, 190
Activator2, 190
active-HIGH, 21
active-LOW, 23
adder, 20
address decoder, 11, 43
Advin Systems, 200
AIM fuse, 18
ALLPRO, 203
Altera, 76, 90, 120, 191
AMD, 67, 190
AND array, 20
AND function, 1
AND-OR/tri-state equivalence, 174
antifuse, 18, 39, 130
antifuse FPGAs, 38
anti-static handling, 166
arithmetic functions, 87
ASICs, 11
  design flow, 16
  migration from FPGA, 169
asynchronous devices
  MACH families, 71
  MAX5000 family, 77
AT6000 family, 116

AT&T Microelectronics, 125, 192
Atmel, 38, 116, 192

Basic Computer Systems, 201
bed of nails test, 15
BEHAVIORAL architecture (VHDL), 51
benchmark circuits, 155
boundary scan testing, 163
BP1200, 201
BP Microsystems, 201
buffer tree, 172
buffers, 14
buried macrocell, 32

cache logic, 38, 119
cache memory, 38, 120
CAD, 55
Cadence Design Systems, 201
CASE syntax, 49
CFB, 90
channel routing, 13
Chiplab, 202
CLB, 104
CLC, 142
clock enable, 175
CMOS
  4000 series, 1
  interface in CPLDs, 30
  structure, 7
command file (for simulation), 60
complement term, 29
component library, 14
conductive tracks, 12
configuring LCAs, 37
counter, 6, 176
CP1128, 201
CPLD, 30
crosstalk, 185
CUPL, 203, 204
custom circuits, 11
customization, 16
Cypress, 101, 139, 193

Index

D-latch, 3, 159
D-type flip-flop, 4
　state equations, 45
　use in PAL, 23
Data I/O, 201
decoder, 4
demultiplexer, 4
Designer, 189
device fitting, 41
diode array, 17
driver (routing channel), 36, 38

EAB, 121
Eclipse, 204
Elan Digital Systems, 202
electron beam deposition, 12
enable flip-flop, 176
entity (VHDL), 51
etching, 11
exclusive-OR function, 2, 43
expander array, 76

fan-out, 14, 172
FastTrack, 80, 120
FFB, 85
FLASH370 family, 101
FLASHlogic, 90
FLEX8000 family, 120
FLEX10k family, 120
flip-flop, 4
　operating speed, 36
floating gate, 18
floor plan, 12
FPLA, 24, 94
FPLS, 26, 98, 179
　complement term, 29
full adder, 20
fuse, 17
FusionPLD, 191

GAL16V8, 24
GAL20V8, 24
GAL22V10, 24
GALs, 24
gate array, 13
gate count, 149, 158
generic macrocell, 23
GLB, 94
granularity (of ASICs and FPGAs), 39
Gray code, 19
ground bounce, 183
GRP, 94

hardware description language, 41, 50
HDFB, 85
hierarchical design, 51

ICT, 97, 194
identity comparator, 43
IF-THEN-ELSE syntax, 49
in-circuit programming, 167
　AT6000 family, 119
　FLASHlogic, 92
　FLEX10k family, 125
　FLEX8000 family, 125
　ispLSI families, 96
　LCAs, 40
　MACH families, 75
　XC2000 family, 116
　XC3000 family, 116
　XC4000 family, 116
　XC5000 family, 116
input cells
　CPLDs, 30
instantiation, 53
integrated circuit manufacturing, 11
Intel, 90
interconnection matrix
　ACT1 family, 130
　ACT2 family, 134
　ACT3 family, 136
　AT6000 family, 116
　CPLDs, 30
　FLASH370 family, 101
　FLASHlogic, 90
　FLEX10k family, 123
　FLEX8000 family, 123
　ispLSI families, 94
　MACH familes, 67
　MAX5000 family, 76
　MAX7000 family, 79
　MAX9000 family, 80
　MPA1000 family, 128
　ORCA, 126
　pASIC380 family, 140
　pLSI families, 94
　QL family, 140
　XC2000 family, 109
　XC3000 family, 110
　XC4000 family, 111
　XC5000 family, 112
　XC7300 family, 85
　XC8100 family, 143
　switching speed, 33
Intergraph, 202

Index

INVERT function, 1
I/O cells
  ACT1 family, 133
  ACT2 family, 134
  ACT3 family, 136
  antifuse FPGAs, 38
  AT6000 family, 119
  CPLDs, 30
  FLASH370 family, 103
  FLASHlogic, 91
  FLEX10k family, 124
  FLEX8000 family, 124
  ispLSI families, 95
  LCAs, 34
  MACH1/MACH2 families, 69
  MACH3/MACH4 families, 74
  MAX5000 family, 78
  MAX7000 family, 80
  MAX9000 family, 81
  MPA1000 family, 129
  ORCA, 127
  pASIC380 family, 140
  pLSI families, 95
  PML, 99
  QL family, 140
  XC2000 family, 109
  XC3000 family, 109
  XC4000 family, 109
  XC5000 family, 109
  XC7300 family, 87
  XC8100 family, 144
ISDATA, 202
ispLSI families, 93

J-K flip-flop, 6
  use in FPLS, 26
JTAG, 47, 163
  ispLSI3000 family, 95
  pLSI3000 family, 95

Karnaugh map, 19

latch, 3
LAB, 76, 79, 80, 120
Lattice Semiconductor, 93, 195
LCA, 34, 104
  configuration, 37
LE, 120
LOG/iC, 203, 204
Logical Devices, 203
logic allocator
  MACH1/MACH2 families, 67
  MACH3/MACH4 families, 72

logic block
  antifuse FPGAs, 38
  CPLDs, 30
  LCAs, 34
logic definition, 41
logic equations, 1, 41, 42
  for state machines, 45
logic gate, 1
  input structures, 9
  symbols, 2
logic operator, 1
logic optimisation, 26, 177
look ahead carry, 178
LSI, 11

MACH families
  MACH1/MACH2, 67
  MACH3/MACH4, 72
MACHXL, 190
macro, 14
macrocell, 23
  ACT1 family, 130
  ACT2 family, 133
  ACT3 family, 135
  asynchronous, 71
  AT6000 family, 118
  buried, 32, 70, 77
  FLASH370 family, 102
  FLASHlogic, 90
  FLEX10k family, 120
  FLEX8000 family, 120
  ispLSI families, 94
  MACH1/MACH2 families, 68
  MACH3/MACH4 families, 73
  MAX5000 family, 77
  MAX7000 family, 79
  MAX9000 family, 80
  MPA1000 family, 128
  ORCA, 125
  PA7140, 97
  pASIC380 family, 139
  pLSI families, 94
  PML, 99
  QL family, 139
  XC2000 family, 104
  XC3000 family, 106
  XC4000 family, 108
  XC5000 family, 108
  XC7300 FFB, 85
  XC7300 HDFB, 86
  XC8100 family, 142
master-slave flip-flop, 4

MAX families
  5000 series, 76
  7000 series, 78
  9000 series, 80
MAX+PLUS, 84, 124, 191
Mealy machine, 48, 98
megablock, 95
Mentor Graphics, 203
metal evaporation, 12
metastability, 182
microprocessor, 11, 146
MicroVia, 142
Minc Inc., 203
modem, 120
Moore machine, 48, 98
MOS switch, 18
Motorola, 128, 196
MPA1000 family, 128
MSI, 4
multiplexer (VHDL), 52

NeoCAD, 203
netlist, 55, 59
NRE costs, 17
operating speed
  logic families, 9

OR array, 24
ORCA, 125
OrCAD Systems, 203
OR function, 1
ORP, 94
output short circuit current, 33

PA7140, 97
PAL16L8, 22
PAL16R4, 23
PAL16R6, 23
PAL16R8, 23
PAL22V10, 24
PALASM, 46, 47, 190
PALs, 19, 42
  registered, 23
parallel logic, 172
partitioning, 146
pASIC380 family, 139
pDS design system, 97, 195

PFU, 125
PGADesigner, 203
Philips, 98, 196
PIA, 76, 79
Pilot-U84, 200
PIM, 102
PIN detector, 26, 43
  logic equations, 45
  schematic, 58
  state equations, 48
  state table, 28
  VHDL definition, 52ff
PIP, 110
PLACE, 98, 194
place and route, 15, 62
PLD, 17
  generic, 24
  security, 37
PLD1100, 201
PLD386+, 204
PLP-S1/S1A, 204
PLS153, 24
PLS155, 26
PLS157, 26
PLS159, 26
PLS173, 24
PLS179, 26
pLSI families, 93
PML, 98
power consumption
  Actel families, 138
  antifuse FPGAs, 39
  AT6000 family, 120
  CPLDs, 32
  FLASH370 family, 103
  FLASHlogic, 92
  FLEX10k family, 125
  FLEX8000 family, 125
  ispLSI families, 96
  LCAs, 36
  logic families, 7
  MACH families, 75
  MAX families, 83
  PA7140, 98
  pASIC380 family, 142
  pLSI families, 96
  QL family, 142
  XC2000 family, 114
  XC3000 families, 115
  XC4000 families, 115
  XC7300 family, 89
  XC8100 family, 145
PREP, 147, 155, 204

Index

primitive cell, 14
product term, 19
  limitation, 24
  savings, 26
  shared, 32
programmable multiplexer, 24
programmable switch, 17, 23
  in LCA routing, 36
programming, 165
  Actel families, 136
  antifuse FPGAs, 40
  FLASH370 family, 103
  FLASHlogic, 92
  ispLSI families, 96
  MACH families, 75
  pLSI families, 96
  XC8100 family, 145
programming time, 40
propagation delay, 147
  Actel families, 138
  antifuse FPGAs, 40
  AT6000 family, 120
  CPLDs, 32
  FLASH370 family, 103
  FLASHlogic, 92
  FLEX10k family, 125
  FLEX8000 family, 125
  ispLSI families, 95
  LCAs, 36
  logic families, 9
  MACH families, 74
  MAX families, 81ff
  PA7140, 98
  pASIC380 family, 142
  pLSI families, 95
  PML, 99
  QL family, 142
  XC2000 family, 113
  XC3000 families, 114
  XC4000 families, 115
  XC5000 families, 115
  XC7300 family, 89
  XC8100 family, 145
PROseries, 205

QL family, 139
Quasar, 204
QuickLogic, 139, 197

RAM blocks, 91, 108, 121, 126
RAM cells, 34, 106

repeater (routing channel), 36, 38
routing channel, 13
  in antifuse FPGAs, 38
  in LCAs, 34
RTL architecture (VHDL), 51

scan path testing, 100
schematic capture, 55
  symbols, 59
Schottky diodes, 9
SDT386+, 204
sea of gates, 13
security, 168
  Actel families, 137
  antifuse FPGAs, 40
  FLASHlogic, 93
  PLDs, 37, 40
sequential circuit, 3
sequential logic, 2
SGUP-85/85A, 205
shared product terms, 32
shift register, 5
signal reflections, 9, 185
silicon dioxide, 11
simulation, 14, 41, 45, 60
  post-layout, 15, 42, 62
  VHDL, 54ff
  waveform generation, 60
slew rate
  limiting, 184
  logic families, 9
SMS Micro Systems, 204
SNAP, 100, 196
SPDE, 141, 197
Sprint Expert, 204
SSI, 4
Stag Programmers, 204
state diagram, 6, 48
  of PIN detector, 27, 44
state equations, 41, 48
state machine, 26, 43, 48, 179
  OHE architecture, 182
  schematic, 56
  VHDL definition, 52ff
Strebor Data, 204
stuck-at-fault analysis, 162
sum of products, 19, 42
sum term, 19
switch matrix
  MACH1/MACH2, 67
  MACH3/MACH4, 72
  MACH3/MACH4 (output), 74

216 Index

Synario, 202
synchronization, 183
Synopsis, 205
synthesis, 50
System General, 205

TangoPLD, 200
testability, 15, 158
test vectors, 47, 62, 158
timing match, 172
timing models
  LSI1016, 96
  MAX5000, 82
  MAX7000, 82
  MAX9000, 83
  XC7300, 89
top-down design, 50, 66, 146
tri-state
  CPLDs, 30
  internal, 174
  LCAs, 35
  PALs, 23
truth table, 1
  for simulation, 46
TTL, 1
  interface in CPLDs, 30
  structure, 7
turbo switch, 33
  MAX families, 83
TURPRO-1, 205

UART, 11
UP2000, 201
UIM, 85
Unisite, 202

validation (of VHDL), 50
vector testing, 47, 158
VeriBest, 202
VersaBlock, 108
VersaRing, 109
VHDL, 50
VHSIC, 50
ViaLink, 140
Viewlogic, 55, 205
VST386+, 204

Warp2/Warp3, 103, 141, 193
wired-OR, 174

XACT, 116, 198
XC2000 family, 104
XC3000 family, 104
XC4000 family, 104
XC5000 family, 104
XC7000 families
  XC7300, 85
  XC7200, 88
XC8100 family, 142
Xilinx, 85, 104, 142, 197